CAMBRIDGE TRACTS IN MATHEMATICS

157 Affine Hecke Algebras and Orthogonal Polynomials

I. G. Macdonald
Queen Mary, University of London

Affine Hecke Algebras and Orthogonal Polynomials

CAMBRIDGE
UNIVERSITY PRESS

PUBLISHED BY THE PRESS SYNDICATE OF THE UNIVERSITY OF CAMBRIDGE
The Pitt Building, Trumpington Street, Cambridge, United Kingdom

CAMBRIDGE UNIVERSITY PRESS
The Edinburgh Building, Cambridge CB2 2RU, UK
40 West 20th Street, New York, NY 10011-4211, USA
477 Williamstown Road, Port Melbourne, VIC 3207, Australia
Ruiz de Alarcón 13, 28014 Madrid, Spain
Dock House, The Waterfront, Cape Town 8001, South Africa

http://www.cambridge.org

First published 2003

Printed in the United Kingdom at the University Press, Cambridge

Typeface Times 10/13 pt *System* LaTeX 2ε [TB]

A catalogue record for this book is available from the British Library

Library of Congress Cataloguing in Publication data

Macdonald, I. G. (Ian Grant)
Affine Hecke algebras and orthogonal polynomials / I.G. Macdonald.
p. cm. – (Cambridge tracts in mathematics; 157)
Includes bibliographical references and index.
ISBN 0 521 82472 9 (hardback)
1. Hecke algebras. 2. Orthogonal polynomials. I. Title. II. Series.
QA174.2 .M28 2003
512′.55 – dc21 2002031075

ISBN 0 521 82472 9 hardback

Contents

Introduction

Over the last fifteen years or so, there has emerged a satisfactory and coherent theory of orthogonal polynomials in several variables, attached to root systems, and depending on two or more parameters. At the present stage of its development, it appears that an appropriate framework for its study is provided by the notion of an affine root system: to each irreducible affine root system S there are associated several families of orthogonal polynomials (denoted by E_λ, P_λ, Q_λ, $P_\lambda^{(\varepsilon)}$ in this book). For example, when S is the non-reduced affine root system of rank 1 denoted here by $(C_1^\vee, C_1)$, the polynomials P_λ are the Askey-Wilson polynomials [A2] which, as is well-known, include as special or limiting cases *all* the classical families of orthogonal polynomials in one variable.

I have surveyed elsewhere [M8] the various antecedents of this theory: symmetric functions, especially Schur functions and their generalizations such as zonal polynomials and Hall-Littlewood functions [M6]; zonal spherical functions on p-adic Lie groups [M1]; the Jacobi polynomials of Heckman and Opdam attached to root systems [H2]; and the constant term conjectures of Dyson, Andrews et al. ([D1], [A1], [M4], [M10]). The lectures of Kirillov [K2] also provide valuable background and form an excellent introduction to the subject.

The title of this monograph is the same as that of the lecture [M7]. That report, for obvious reasons of time and space, gave only a cursory and incomplete overview of the theory. The modest aim of the present volume is to fill in the gaps in that report and to provide a unified foundation for the theory in its present state.

The decision to treat all affine root systems, reduced or not, simultaneously on the same footing has resulted in an unavoidably complex system of notation. In order to formulate results uniformly it is necessary to associate to each affine root system S another affine root system S' (which may or may not coincide with S), and to each labelling (§1.5) of S a dual labelling of S'.

The prospective reader is expected to be familiar with the algebra and geometry of (crystallographic) root systems and Weyl groups, as expounded for example by Bourbaki in [B1]. Beyond that, the book is pretty well self-contained.

We shall now survey briefly the various chapters and their contents. The first four chapters are preparatory to Chapter 5, which contains all the main results. Chapter 1 covers the basic properties of affine root systems and their classification. Chapter 2 is devoted to the extended affine Weyl group, and collects various notions and results that will be needed later.

Chapter 3 introduces the (Artin) braid group of an extended affine Weyl group, and the double braid group. The main result of this chapter is the duality theorem (3.5.1); although it is fundamental to the theory, there is at this time of writing no complete proof in the literature. I have to confess that the proof given here of the duality theorem is the least satisfactory feature of the book, since it consists in checking, in rather tedious detail, the necessary relations between the generators. Fortunately, B. Ion [I1] has recently given a more conceptual proof which avoids these calculations.

The subject of Chapter 4 is the affine Hecke algebra $\mathfrak{H}$, which is a deformation of the group algebra of the extended affine Weyl group. We construct the basic representation of $\mathfrak{H}$ in §4.3 and develop its properties in the subsequent sections. Finally, in §4.7 we introduce the double affine Hecke algebra $\tilde{\mathfrak{H}}$, and show that the duality theorem for the double braid group gives rise to a duality theorem for $\tilde{\mathfrak{H}}$.

As stated above, Chapter 5, on orthogonal polynomials, is the heart of the book. The scalar products are introduced in §5.1, the orthogonal polynomials E_λ in §5.2, the symmetric orthogonal polynomials P_λ in §5.3, and their variants Q_λ and $P_\lambda^{(\varepsilon)}$ in §5.7. The main results of the chapter are the symmetry theorems (5.2.4) and (5.3.5); the specialization theorems (5.2.14) and (5.3.12); and the norm formulas (5.8.17) and (5.8.19), which include as special cases almost all the constant term conjectures referred to earlier.

The final Chapter 6 deals with the case where the affine root system S has rank 1. Here everything can be made completely explicit. When S is of type A_1, the polynomials P_λ are the continuous q-ultraspherical (or Rogers) polynomials, and when S is of type $(C_1^\vee, C_1)$ they are the Askey-Wilson polynomials, as mentioned above.

The subject of this monograph has many connections with other parts of mathematics and theoretical physics, such as (in no particular order) algebraic combinatorics, harmonic analysis, integrable quantum systems, quantum groups and symmetric spaces, quantum statistical mechanics, deformed

Virasoro algebras, and string theory. I have made no attempt to survey these various applications, partly from lack of competence, but also because an adequate account would require a book of its own.

Finally, references to the history and the literature will be found in the Notes and References at the end of each chapter.

1

Affine root systems

1.1 Notation and terminology

Let E be an affine space over a field K: that is to say, E is a set on which a K-vector space V acts faithfully and transitively. The elements of V are called *translations* of E, and the effect of a translation $v \in V$ on $x \in E$ is written $x + v$. If $y = x + v$ we write $v = y - x$.

Let E' be another affine space over K, and let V' be its vector space of translations. A mapping $f : E \to E'$ is said to be *affine-linear* if there exists a K-linear mapping $Df : V \to V'$, called the *derivative* of f, such that

$$f(x + v) = f(x) + (Df)(v). \tag{1.1.1}$$

for all $x \in E$ and $v \in V$. In particular, a function $f\colon E \to K$ is affine-linear if and only if there exists a linear form $Df\colon V \to K$ such that (1.1.1) holds.

If $f, g : E \to K$ are affine-linear and $\lambda, \mu \in K$, the function $h = \lambda f + \mu g : x \mapsto \lambda f(x) + \mu g(x)$ is affine-linear, with derivative $Dh = \lambda Df + \mu Dg$. Hence the set F of all affine-linear functions $f\colon E \to K$ is a K-vector space, and D is a K-linear mapping of F onto the dual V^* of the vector space V. The kernel of D is the 1-dimensional subspace F^0 of F consisting of the constant functions.

Let F^* be the dual of the vector space F. For each $x \in E$, the evaluation map $\varepsilon_x : f \mapsto f(x)$ belongs to F^*, and the mapping $x \mapsto \varepsilon_x$ embeds E in F^* as an affine hyperplane. Likewise, for each $v \in V$ let $\varepsilon_v \in F^*$ be the mapping $f \mapsto (Df)(v)$. If $v = y - x$, where $x, y \in E$, we have $\varepsilon_v = \varepsilon_y - \varepsilon_x$ by (1.1.1), and the mapping $v \mapsto \varepsilon_v$ embeds V in F^* as the hyperplane through the origin parallel to E.

From now on, K will be the field $\mathbb{R}$ of real numbers, and V will be a real vector space of finite dimension $n > 0$, equipped with a positive definite symmetric scalar product $<u, v>$. We shall write

$$|v| = <v, v>^{1/2}$$

for the length of a vector $v \in V$. Then E is a Euclidean space of dimension n, and is a metric space for the distance function $d(x, y) = |x - y|$.

We shall identify V with its dual space V^* by means of the scalar product $<u, v>$. For any affine-linear function $f: E \to \mathbb{R}$, (1.1.1) now takes the form

(1.1.2) $$f(x + v) = f(x) + <Df, v>$$

and Df is the *gradient* of f, in the usual sense of calculus.

We define a scalar product on the space F as follows:

(1.1.3) $$<f, g> = <Df, Dg>.$$

This scalar product is positive semidefinite, with radical the one-dimensional space F^0 of constant functions.

For each $v \neq 0$ in V let

$$v^\vee = 2v/|v|^2$$

and for each non-constant $f \in F$ let

$$f^\vee = 2f/|f|^2.$$

Also let

$$H_f = f^{-1}(0)$$

which is an affine hyperplane in E. The reflection in this hyperplane is the isometry $s_f: E \to E$ given by the formula

(1.1.4) $$s_f(x) = x - f^\vee(x)Df = x - f(x)Df^\vee.$$

By transposition, s_f acts on F: $s_f(g) = g \circ s_f^{-1} = g \circ s_f$. Explicitly, we have

(1.1.5) $$s_f(g) = g - <f^\vee, g>f = g - <f, g>f^\vee$$

for $g \in F$.

For each $u \neq 0$ in V, let $s_u: V \to V$ denote the reflection in the hyperplane orthogonal to u, so that

(1.1.6) $$s_u(v) = v - <u, v>u^\vee.$$

Then it is easily checked that

(1.1.7) $$Ds_f = s_{Df}$$

for any non constant $f \in F$.

Let $w : E \to E$ be an isometry. Then w is affine-linear (because it preserves parallelograms) and its derivative Dw is a linear isometry of V, i.e., we have $<(Dw)u, (Dw)v> = <u, v>$ for all $u, v \in V$. The mapping w acts by transposition on F: $(wf)(x) = f(w^{-1}x)$ for $x \in V$, and we have

(1.1.8) $$D(wf) = (Dw)(Df).$$

For each $v \in V$ we shall denote by $t(v) : E \to E$ the translation by v, so that $t(v)x = x + v$. The translations are the isometries of E whose derivative is the identity mapping of V. On F, $t(v)$ acts as follows:

(1.1.9) $$t(v)f = f - <Df, v>c$$

where c is the constant function equal to 1. For if $x \in E$ we have

$$(t(v)f)(x) = f(x - v) = f(x) - <Df, v>.$$

Let w: $E \to E$ be an isometry and let $v \in V$. Then

(1.1.10) $$wt(v)w^{-1} = t((Dw)v).$$

For if $x \in E$ we have

$$(wt(v)w^{-1})(x) = w(w^{-1}x + v) = x + (Dw)v.$$

1.2 Affine root systems

As in §1.1 let E be a real Euclidean space of dimension $n > 0$, and let V be its vector space of translations. We give E the usual topology, defined by the metric $d(x, y) = |x - y|$, so that E is locally compact. As before, let F denote the space (of dimension $n + 1$) of affine-linear functions on E.

An *affine root system* on E [M2] is a subset S of F satisfying the following axioms (AR1)–(AR4):

(AR 1) S spans F, *and the elements of* S *are non-constant functions.*
(AR 2) $s_a(b) \in S$ *for all* $a, b \in S$.
(AR 3) $<a^\vee, b> \in \mathbb{Z}$ *for all* $a, b \in S$.

The elements of S are called *affine roots*, or just *roots*. Let W_S be the group of isometries of E generated by the reflections s_a for all $a \in S$. This group W_S is the *Weyl group* of S. The fourth axiom is now

(AR 4) *W_S (as a discrete group) acts properly on E.*

In other words, if K_1 and K_2 are compact subsets of E, the set of $w \in W_S$ such that $wK_1 \cap K_2 \neq \emptyset$ is *finite*.

From (AR3) it follows, just as in the case of a finite root system, that if a and λa are proportional affine roots, then λ is one of the numbers $\pm\frac{1}{2}, \pm 1, \pm 2$. If $a \in S$ and $\frac{1}{2}a \notin S$, the root a is said to be *indivisible*. If each $a \in S$ is indivisible, i.e., if the only roots proportional to $a \in S$ are $\pm a$, the root system S is said to be *reduced*.

If S is an affine root system on E, then

$$S^\vee = \{a^\vee : a \in S\}$$

is also an affine root system on E, called the *dual* of S. Clearly S and $S^\vee$ have the same Weyl group, and $S^{\vee\vee} = S$.

The *rank* of S is defined to be the dimension n of E (or V). If S' is another affine root system on a Euclidean space E', an *isomorphism* of S onto S' is a bijection of S onto S' that is induced by an isometry of E onto E'. If S' is isomorphic to λS for some nonzero $\lambda \in \mathbb{R}$, we say that S and S' are *similar*.

We shall assume throughout that S is *irreducible*, i.e. that there exists no partition of S into two non-empty subsets S_1, S_2 such that $<a_1, a_2> = 0$ for all $a_1 \in S_1$ and $a_2 \in S_2$.

The following proposition ([M2], p. 98) provides examples of affine root systems:

(1.2.1) *Let R be an irreducible finite root system spanning a real finite-dimensional vector space V, and let $<u, v>$ be a positive-definite symmetric bilinear form on V, invariant under the Weyl group of R. For each $\alpha \in R$ and $r \in \mathbb{Z}$ let $a_{\alpha,r}$ denote the affine-linear function on V defined by*

$$a_{\alpha,r}(x) = <\alpha, x> + r.$$

Then the set $S(R)$ of functions $a_{\alpha,r}$, where $\alpha \in R$ and r is any integer if $\frac{1}{2}\alpha \notin R$ (resp. any odd integer if $\frac{1}{2}\alpha \in R$) is a reduced irreducible affine root system on V.

Moreover, every reduced irreducible affine root system is similar to either $S(R)$ or $S(R)^\vee$, where R is a finite (but not necessarily reduced) irreducible root system ([M2], §6).

Let S be an irreducible affine root system on a Euclidean space E. The set $\{H_a \colon a \in S\}$ of affine hyperplanes in E on which the affine roots vanish is locally finite ([M2], §4). Hence the set $E - \bigcup_{a\in S} H_a$ is open in E, and therefore so also are the connected components of this set, since E is locally connected. These components are called the *alcoves* of S, or of W_S, and it is a basic fact (loc. cit.) that the Weyl group W_S acts faithfully and transitively on the set of alcoves. Each alcove is an open rectilinear n-simplex, where n is the rank of S.

Choose an alcove C once and for all. Let x_i $(i\in I)$ be the vertices of C, so that C is the set of all points $x = \sum \lambda_i x_i$ such that $\sum \lambda_i = 1$ and each λ_i is a positive real number. Let $B = B(C)$ be the set of indivisible affine roots $a \in S$ such that (i) H_a is a wall of C, and (ii) $a(x) > 0$ for all $x \in C$. Then B consists of $n + 1$ roots, one for each wall of C, and B is a basis of the space F of affine-linear functions on E. The set B is called a *basis* of S.

The elements of B will be denoted by a_i $(i\in I)$, the notation being chosen so that $a_i(x_j) = 0$ if $i \neq j$. Since x_i is in the closure of C, we have $a_i(x_i) > 0$. Moreover, $<a_i, a_j> \leq 0$ whenever $i \neq j$.

The alcove C having been chosen, an affine root $a \in S$ is said to be *positive* (resp. *negative*) if $a(x) > 0$ (resp. $a(x) < 0$) for $x \in C$. Let S^+ (resp. S^-) denote the set of positive (resp. negative) affine roots; then $S = S^+ \cup S^-$ and $S^- = -S^+$. Moreover, each $a \in S^+$ is a linear combination of the a_i with nonnegative integer coefficients, just as in the finite case ([M2], §4).

Let $\alpha_i = Da_i$ $(i\in I)$. The $n + 1$ vectors $\alpha_i \in V$ are linearly dependent, since $\dim V = n$. There is a unique linear relation of the form

$$\sum_{i\in I} m_i \alpha_i = 0$$

where the m_i are positive integers with no common factor, and at least one of the m_i is equal to 1. Hence the function

$$c = \sum_{i\in I} m_i a_i \tag{1.2.2}$$

is constant on E (because its derivative is zero) and positive (because it is positive on C).

Let

$$\Sigma = \{Da : a \in S\}.$$

Then Σ is an irreducible (finite) root system in V. A vertex x_i of the alcove C is said to be *special* for S if (i) $m_i = 1$ and (ii) the vectors α_j ($j \in I$, $j \neq i$) form a basis of Σ. For each affine root system S there is at least one special vertex (see the tables in §1.3). We shall choose a special vertex once and for all, and denote it by x_0 (so that 0 is a distinguished element of the index set I). Thus $m_0 = 1$ in (1.2.2), and if we take x_0 as origin in E, thereby identifying E with V, the affine root a_i ($i \neq 0$) is identified with α_i.

The Cartan matrix and the Dynkin diagram of an irreducible affine root system S are defined exactly as in the finite case. The *Cartan matrix* of S is the matrix $N = (n_{ij})_{i,j \in I}$ where $n_{ij} = \langle a_i^\vee, a_j \rangle$. It has $n + 1$ rows and columns, and its rank is n. Its diagonal entries are all equal to 2, and its off-diagonal entries are integers ≤ 0. If $m = (m_i)_{i \in I}$ is the column vector formed by the coefficients in (1.2.2), we have $Nm = 0$.

The *Dynkin diagram* of S is the graph with vertex set I, in which each pair of distinct vertices i, j is joined by d_{ij} edges, where $d_{ij} = \max(|n_{ij}|, |n_{ji}|)$. We have $d_{ij} \leq 4$ in all cases. For each pair of vertices i, j such that $d_{ij} > 0$ and $|a_i| > |a_j|$, we insert an arrowhead (or inequality sign) pointing towards the vertex j corresponding to the shorter root.

If S is reduced, the Dynkin diagram of $S^\vee$ is obtained from that of S by reversing all arrowheads. If $S = S(R)$ as in (1.2.1), where R is irreducible and reduced, the Dynkin diagram of S is the 'completed Dynkin diagram' of R([B1], ch. 6).

If S is reduced, the Cartan matrix and the Dynkin diagram each determine S up to similarity. If S is not reduced, the Dynkin diagram still determines S, provided that the vertices $i \in I$ such that $2a_i \in S$ are marked (e.g. with an asterisk).

1.3 Classification of affine root systems

Let S be an irreducible affine root system. If S is reduced, then S is similar to either $S(R)$ or $S(R)^\vee$ (1.2.1), where R is an irreducible root system. If R is of type X, where X is one of the symbols A_n, B_n, C_n, D_n, BC_n, E_6, E_7, E_8, F_4, G_2, we say that $S(R)$ (resp. $S(R)^\vee$) is of type X (resp. $X^\vee$).

If S is not reduced, it determines two reduced affine root systems

$$S_1 = \{a \in S : \tfrac{1}{2}a \notin S\}, \quad S_2 = \{a \in S : 2a \notin S\}$$

with the same affine Weyl group, and $S = S_1 \cup S_2$. We say that S is of type (X, Y) where X, Y are the types of S_1, S_2 respectively.

The reduced and non-reduced irreducible affine root systems are listed below ((1.3.1)–(1.3.18)). In this list, $\varepsilon_1, \varepsilon_2, \dots$ is a sequence of orthonormal vectors in a real Hilbert space.

For each type we shall exhibit

(a) an affine root system S of that type;
(b) a basis of S;
(c) the Dynkin diagram of S. Here the numbers attached to the vertices of the diagram are the coefficients m_i in (1.2.2).

We shall first list the reduced systems ((1.3.1)–(1.3.14)) and then the non-reduced systems ((1.3.15)–(1.3.18)).

(1.3.1) *Type* A_n $(n \geq 1)$.

(a) $\pm(\varepsilon_i - \varepsilon_j) + r$ $(1 \leq i < j \leq n+1;\ r \in \mathbb{Z})$.
(b) $a_0 = -\varepsilon_1 + \varepsilon_{n+1} + 1,\quad a_i = \varepsilon_i - \varepsilon_{i+1}$ $(1 \leq i \leq n)$.

(c) 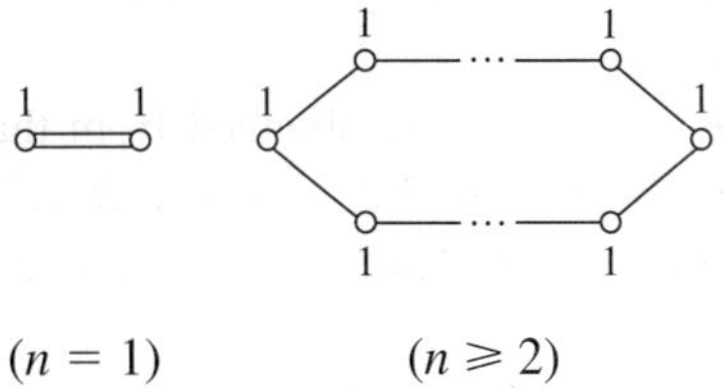

(1.3.2) *Type* B_n $(n \geq 3)$.

(a) $\pm\varepsilon_i + r$ $(1 \leq i \leq n;\ r \in \mathbb{Z})$; $\pm\varepsilon_i \pm \varepsilon_j + r$ $(1 \leq i < j \leq n;\ r \in \mathbb{Z})$.
(b) $a_0 = -\varepsilon_1 - \varepsilon_2 + 1,\quad a_i = \varepsilon_i - \varepsilon_{i+1}$ $(1 \leq i \leq n-1),\quad a_n = \varepsilon_n$.

(c) 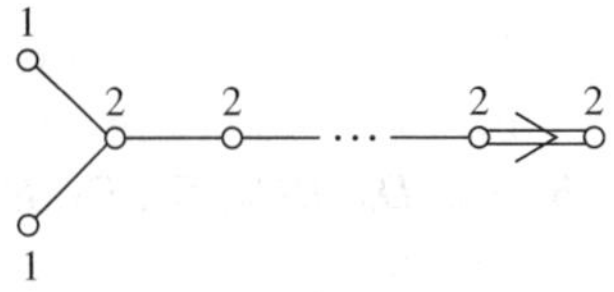

(1.3.3) *Type* $B_n^\vee$ $(n \geq 3)$.

(a) $\pm 2\varepsilon_i + 2r$ $(1 \leq i \leq n;\ r \in \mathbb{Z})$; $\pm\varepsilon_i \pm \varepsilon_j + r$ $(1 \leq i < j \leq n; r \in \mathbb{Z})$.

(b) $a_0 = -\varepsilon_1 - \varepsilon_2 + 1, \quad a_i = \varepsilon_i - \varepsilon_{i+1}\ (1 \le i \le n-1), \quad a_n = 2\varepsilon_n.$

(c) 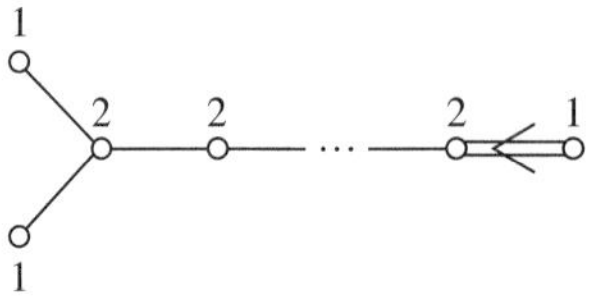

(1.3.4) *Type C_n ($n \ge 2$).*

(a) $\pm 2\varepsilon_i + r\ (1 \le i \le n;\ r \in \mathbb{Z}); \quad \pm\varepsilon_i \pm \varepsilon_j + r\ (1 \le i < j \le n; r \in \mathbb{Z}).$

(b) $a_0 = -2\varepsilon_1 + 1, \quad a_i = \varepsilon_i - \varepsilon_{i+1}\ (1 \le i \le n-1), \quad a_n = 2\varepsilon_n.$

(c) 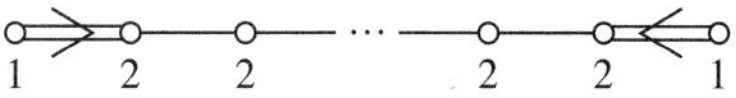

(1.3.5) *Type $C_n^\vee$ ($n \ge 2$).*

(a) $\pm\varepsilon_i + \frac{1}{2}r\ (1 \le i \le n;\ r \in \mathbb{Z}); \quad \pm\varepsilon_i \pm \varepsilon_j + r\ (1 \le i < j \le n;\ r \in \mathbb{Z}).$

(b) $a_0 = -\varepsilon_1 + \frac{1}{2}, \quad a_i = \varepsilon_i - \varepsilon_{i+1}\ (1 \le i \le n-1), \quad a_n = \varepsilon_n.$

(c) 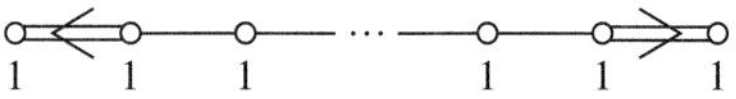

(1.3.6) *Type BC_n ($n \ge 1$).*

(a) $\pm\varepsilon_i + r\ (1 \le i \le n;\ r \in \mathbb{Z}); \quad \pm 2\varepsilon_i + 2r + 1\ (1 \le i \le n;\ r \in \mathbb{Z});$

$$\pm\varepsilon_i \pm \varepsilon_j + r\ (1 \le i < j \le n;\ r \in \mathbb{Z}).$$

(b) $a_0 = -2\varepsilon_1 + 1, \quad a_i = \varepsilon_i - \varepsilon_{i+1}\ (1 \le i \le n-1), \quad a_n = \varepsilon_n.$

(c) 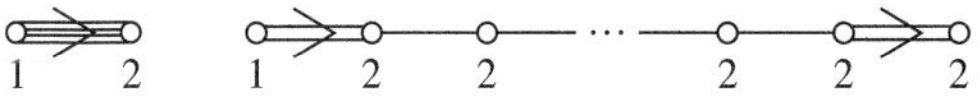

($n = 1$) ($n \geqslant 2$)

(1.3.7) *Type D_n ($n \ge 4$).*

(a) $\pm\varepsilon_i \pm \varepsilon_j + r\ (1 \le i < j \le n;\ r \in \mathbb{Z})$

(b) $a_0 = -\varepsilon_1 - \varepsilon_2 + 1, \quad a_i = \varepsilon_i - \varepsilon_{i+1}\ (1 \le i \le n-1), \quad a_n = \varepsilon_{n-1} + \varepsilon_n.$

(c) 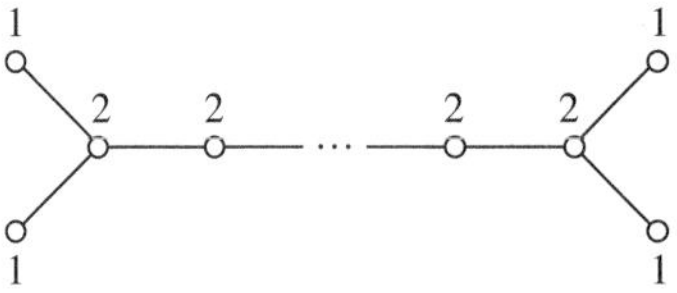

These are the "classical" reduced affine root systems. The next seven types ((1.3.8)–(1.3.14)) are the "exceptional" reduced affine root systems. In (1.3.8)–(1.3.10) let

$$\omega_i = \varepsilon_i - \frac{1}{9}(\varepsilon_1 + \cdots + \varepsilon_9) \qquad (1 \le i \le 9).$$

(1.3.8) *Type* E_6.

(a) $\pm(\omega_i - \omega_j) + r$ $(1 \le i < j \le 6;\ r \in \mathbb{Z})$;
$\pm(\omega_i + \omega_j + \omega_k) + r$ $(1 \le i < j < k \le 6;\ r \in \mathbb{Z})$;
$\pm(\omega_i + \omega_2 + \cdots + \omega_6) + r$ $(r \in \mathbb{Z})$.

(b) $a_0 = -(\omega_1 + \cdots + \omega_6) + 1$, $a_i = \omega_i - \omega_{i+1}$ $(1 \le i \le 5)$,
$a_6 = \omega_4 + \omega_5 + \omega_6$.

(c) 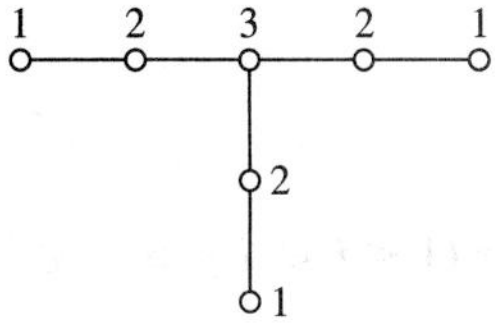

(1.3.9) *Type* E_7.

(a) $\pm(\omega_i - \omega_j) + r$ $(1 \le i < j \le 7;\ r \in \mathbb{Z})$;
$\pm(\omega_i + \omega_j + \omega_k) + r$ $(1 \le i < j < k \le 7;\ r \in \mathbb{Z})$;
$\pm(\omega_1 + \cdots + \hat{\omega}_i + \cdots + \omega_7) + r$ $(1 \le i \le 7;\ r \in \mathbb{Z})$.

(b) $a_0 = -(\omega_1 + \cdots + \omega_6) + 1$, $a_i = \omega_i - \omega_{i+1}$ $(1 \le i \le 6)$,
$a_7 = \omega_5 + \omega_6 + \omega_7$.

(c) 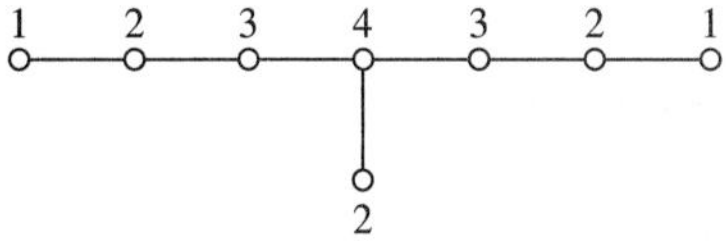

(1.3.10) *Type* E_8.

(a) $\pm(\omega_i - \omega_j) + r$ $(1 \le i < j \le 9;\ r \in \mathbb{Z})$;
$\pm(\omega_i + \omega_j + \omega_k) + r$ $(1 \le i < j < k \le 9;\ r \in \mathbb{Z})$.

(b) $a_0 = \omega_1 - \omega_2 + 1$, $a_i = \omega_{i+1} - \omega_{i+2}$ $(1 \le i \le 7)$,
$a_8 = \omega_7 + \omega_8 + \omega_9$.

(c) 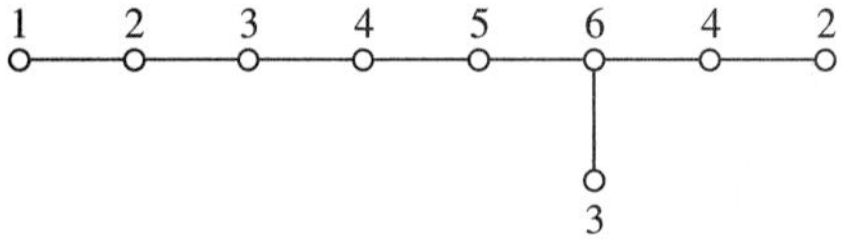

(1.3.11) *Type* F_4.

(a) $\pm\varepsilon_i + r\ (1 \le i \le 4;\ r \in \mathbb{Z})$; $\pm\varepsilon_i \pm \varepsilon_j + r\ (1 \le i < j \le 4;\ r \in \mathbb{Z})$;
$\frac{1}{2}(\pm\varepsilon_1 \pm \varepsilon_2 \pm \varepsilon_3 \pm \varepsilon_4) + r\ (r \in \mathbb{Z})$.

(b) $a_0 = -\varepsilon_1 - \varepsilon_2 + 1,\quad a_1 = \varepsilon_2 - \varepsilon_3,\quad a_2 = \varepsilon_3 - \varepsilon_4,\quad a_3 = \varepsilon_4,$
$a_4 = \frac{1}{2}(\varepsilon_1 - \varepsilon_2 - \varepsilon_3 - \varepsilon_4)$.

(c) 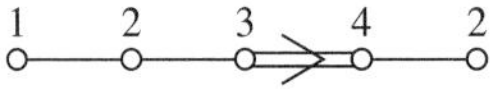

(1.3.12) *Type* $F_4^\vee$.

(a) $\pm 2\varepsilon_i + 2r\ (1 \le i \le 4;\ r \in \mathbb{Z})$; $\pm\varepsilon_i \pm \varepsilon_j + r\ (1 \le i < j \le 4;\ r \in \mathbb{Z})$;
$\pm\varepsilon_1 \pm \varepsilon_2 \pm \varepsilon_3 \pm \varepsilon_4 + 2r\ (r \in \mathbb{Z})$.

(b) $a_0 = -\varepsilon_1 - \varepsilon_2 + 1,\quad a_1 = \varepsilon_2 - \varepsilon_3,\quad a_2 = \varepsilon_3 - \varepsilon_4,\quad a_3 = 2\varepsilon_4,$
$a_4 = \varepsilon_1 - \varepsilon_2 - \varepsilon_3 - \varepsilon_4$.

(c) 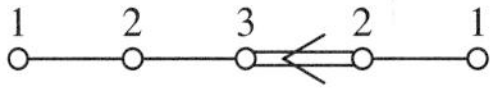

(1.3.13) *Type* G_2.

(a) $\pm(\varepsilon_i - \frac{1}{3}(\varepsilon_1 + \varepsilon_2 + \varepsilon_3)) + r\ (1 \le i \le 3;\ r \in \mathbb{Z})$;
$\pm(\varepsilon_i - \varepsilon_j) + r\ (1 \le i < j \le 3;\ r \in \mathbb{Z})$.

(b) $a_0 = \varepsilon_1 - \varepsilon_2 + 1,\quad a_1 = \varepsilon_2 - \varepsilon_3,\quad a_2 = \varepsilon_3 - \frac{1}{3}(\varepsilon_1 + \varepsilon_2 + \varepsilon_3)$.

(c) 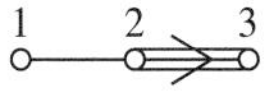

(1.3.14) *Type* $G_2^\vee$.

(a) $\pm(3\varepsilon_i - (\varepsilon_1 + \varepsilon_2 + \varepsilon_3)) + 3r\ (1 \le i \le 3;\ r \in \mathbb{Z})$;
$\pm(\varepsilon_i - \varepsilon_j) + r\ (1 \le i < j \le 3;\ r \in \mathbb{Z})$.

(b) $a_0 = \varepsilon_1 - \varepsilon_2 + 1,\quad a_1 = \varepsilon_2 - \varepsilon_3,\quad a_2 = 3\varepsilon_3 - (\varepsilon_1 + \varepsilon_2 + \varepsilon_3)$.

(c) 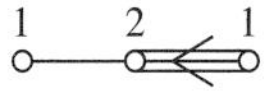

We come now to the non-reduced affine root systems. In the Dynkin diagrams below, an asterisk placed over a vertex indicates that if a_i is the affine root corresponding to that vertex in a basis of S, then $2a_i \in S$.

(1.3.15) *Type* (BC_n, C_n) $(n \ge 1)$.

(a) $\pm\varepsilon_i + r,\quad \pm 2\varepsilon_i + r\ (1 \le i \le n, r \in \mathbb{Z})$;
$\pm\varepsilon_i \pm \varepsilon_j + r\ (1 \le i < j \le n;\ r \in \mathbb{Z})$.

(b) $a_0 = -2\varepsilon_1 + 1, \quad a_i = \varepsilon_i - \varepsilon_{i+1}\ (1 \le i \le n-1), \quad a_n = \varepsilon_n.$

(c) 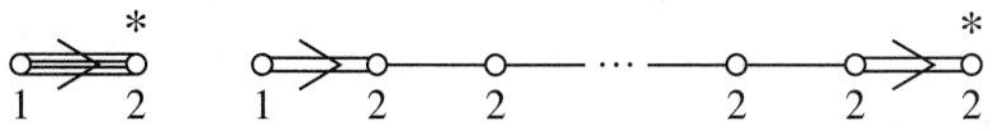

$(n = 1)$ $(n \geqslant 2)$

(1.3.16) *Type* $(C_n^\vee, BC_n)$ $(n \ge 1)$.

(a) $\pm\varepsilon_i + \frac{1}{2}r, \quad \pm 2\varepsilon_i + 2r\ (1 \le i \le n;\ r \in \mathbb{Z});$
$\pm\varepsilon_i \pm \varepsilon_j + r\ (1 \le i < j \le n;\ r \in \mathbb{Z}).$

(b) $a_0 = -\varepsilon_1 + \frac{1}{2}, \quad a_i = \varepsilon_i - \varepsilon_{i+1}, \quad a_n = \varepsilon_n.$

(c) 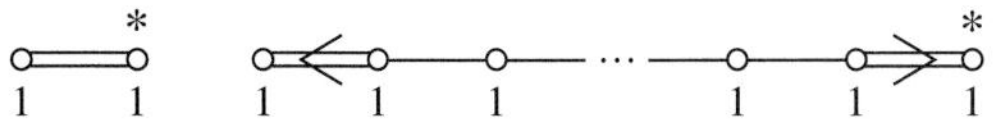

$(n = 1)$ $(n \geqslant 2)$

(1.3.17) *Type* $(C_2, C_2^\vee), (B_n, B_n^\vee)$ $(n \ge 3)$.

(a) $\pm\varepsilon_i + r, \quad \pm 2\varepsilon_i + 2r\ (1 \le i \le n;\ r \in \mathbb{Z});$
$\pm\varepsilon_i \pm \varepsilon_j + r\ (1 \le i < j \le n;\ r \in \mathbb{Z}).$

(b) $a_0 = -\varepsilon_1 - \varepsilon_2 + 1, \quad a_i = \varepsilon_i - \varepsilon_{i+1}\ (1 \le i \le n-1); \quad a_n = \varepsilon_n.$

(c) 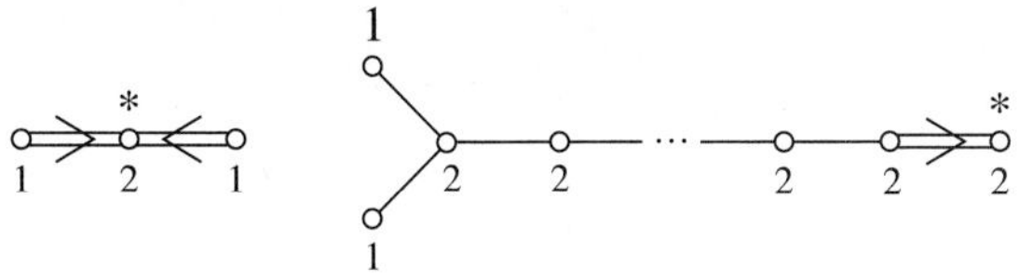

$(n = 2)$ $(n \geqslant 3)$

(1.3.18) *Type* $(C_n^\vee, C_n)$ $(n \ge 1)$.

(a) $\pm\varepsilon_i + \frac{1}{2}r, \quad \pm 2\varepsilon_i + r\ (1 \le i \le n;\ r \in \mathbb{Z});$
$\pm\varepsilon_i \pm \varepsilon_j + r\ (1 \le i < j \le n;\ r \in \mathbb{Z}).$

(b) $a_0 = -\varepsilon_1 + \frac{1}{2}, \quad a_i = \varepsilon_i - \varepsilon_{i+1}\ (1 \le i \le n-1), \quad a_n = \varepsilon_n.$

(c) 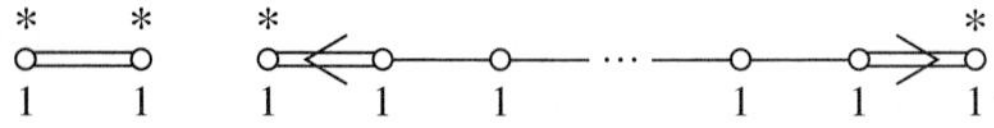

$(n = 1)$ $(n \geqslant 2)$

For each irreducible affine root system S, let $o(S)$ denote the number of W_S-orbits in S. If S is reduced, the list above shows that $o(S) \le 3$, and that $o(S) = 3$

only when S is of type C_n, $C_n^\vee$ or $BC_n (n \geq 2)$. If S is not reduced, the maximum value of $o(S)$ is 5, and is attained only when S is of type $(C_n^\vee, C_n)$ $(n \geq 2)$. The five orbits are $O_1, \ldots, O_5$ where, in the notation of (1.3.18) above,

$$O_1 = \{\pm\varepsilon_i + r : 1 \leq i \leq n, r \in \mathbb{Z}\}, \quad O_2 = 2O_1, \quad O_3 = O_1 + \tfrac{1}{2},$$
$$O_4 = 2O_3 = O_2 + 1, \quad O_5 = \{\pm\varepsilon_i \pm \varepsilon_j + r : 1 \leq i < j \leq n,\ r \in \mathbb{Z}\}.$$

Finally, the list above shows that all the non-reduced irreducible affine root systems of rank n are subsystems of (1.3.18), obtained by deleting one or more of the W_S-orbits; and so are the "classical" root systems (1.3.2)–(1.3.7).

1.4 Duality

In later chapters, in order to formulate conveniently certain dualities, we shall need to consider not one but a pair (S, S') of irreducible affine root systems, together with a pair (R, R') of finite root systems and a pair (L, L') of lattices in V.

Let R be a reduced finite irreducible root system in V, and let P (resp. $P^\vee$) denote the weight lattice of R (resp. $R^\vee$), and Q (resp. $Q^\vee$) the root lattice of R (resp. $R^\vee$). Fix a basis $(\alpha_i)_{i \in I_0}$ of R, and let φ be the highest root of R relative to this basis. In (1.4.1) and (1.4.2) below we shall assume that the scalar product on V is normalized so that $|\varphi|^2 = 2$ and therefore $\varphi^\vee = \varphi$. (This conflicts with standard usage, as in § 1.3, only when R is of type C_n (1.3.4).)

The pairs (S, S'), (R, R'), (L, L') to be considered are the following:

(1.4.1) $\quad S = S(R), \quad S' = S(R^\vee); \quad R' = R^\vee; \quad L = P, \quad L' = P^\vee.$

Then S (resp. S') has a basis $(a_i)_{i \in I}$ (resp. $(a'_i)_{i \in I}$) where $a_i = \alpha_i$ $(i \neq 0)$, $a_0 = -\varphi + c$; $a'_i = \alpha_i^\vee$ $(i \neq 0)$, $a'_0 = -\psi^\vee + c$, where ψ is the highest *short* root of R.

(1.4.2) $\qquad S = S' = S(R)^\vee; \quad R' = R; \quad L = L' = P^\vee.$

Then $S = S'$ has a basis $(a_i)_{i \in I} = (a'_i)_{i \in I}$, where $a_i = a'_i = \alpha_i^\vee$ if $i \neq 0$, and $a_0 = a'_0 = -\varphi + c$.

(1.4.3) $\quad S = S'$ is of type $(C_n^\vee, C_n)$; $R = R'$ is of type C_n; $L = L' = Q^\vee$. We shall assume that S is as given in (1.3.18), so that $a_i = \alpha_i = \varepsilon_i - \varepsilon_{i+1}$ $(1 \leq i \leq n-1)$ and $\alpha_n = 2a_n = 2\varepsilon_n$, and $L = \mathbb{Z}^n$.

For each $\alpha \in R$, let $\alpha'(= \alpha$ or $\alpha^\vee)$ be the corresponding element of R'. Then $\langle\lambda', \alpha\rangle$ and $\langle\lambda, \alpha'\rangle$ are integers, for all $\lambda \in L$, $\lambda' \in L'$ and $\alpha \in R$.

In each case let

(1.4.4) $$\Omega' = L'/Q^\vee,$$

a finite abelian group. Also let

$$<L, L'> = \{<\lambda, \lambda'> : \lambda \in L, \lambda' \in L'\}.$$

Then we have

(1.4.5) $$<L, L'> = e^{-1}\mathbb{Z}$$

where e is the exponent of Ω', except in case (1.4.2) when R is of type B_n or C_{2n}, in which case $e = 1$.

Anticipating Chapter 2, let $W = W(R, L')$ be the group of displacements of V generated by the Weyl group W_0 of R and the translations $t(\lambda')$, $\lambda' \in L'$, so that W is the semidirect product of W_0 and $t(L')$:

(1.4.6) $$W = W(R, L') = W_0 \ltimes t(L').$$

Dually, let

(1.4.6′) $$W' = W(R', L) = W_0 \ltimes t(L).$$

By transposition, both W and W' act on F.

(1.4.7) *W permutes S and W' permutes S'.*

This follows from the fact, remarked above, that $<\lambda', \alpha>$ and $<\lambda, \alpha'>$ are integers, for all $\lambda \in L$, $\lambda' \in L'$ and $\alpha \in R$.

Now let

(1.4.8) $$\Lambda = L \oplus \mathbb{Z}c_0$$

where $c_0 = e^{-1}c$. We shall regard elements of Λ as functions on V: if $f \in \Lambda$, say $f = \lambda + rc_0$ where $\lambda \in L$ and $r \in \mathbb{Z}$, then

$$f(x) = <\lambda, x> + e^{-1}r$$

for $x \in V$. Then Λ is a lattice in F.

(1.4.9) *Λ is stable under the action of W.*

Proof Let $w \in W$, say $w = vt(\lambda')$ where $v \in W_0$ and $\lambda' \in L'$. If $f = \lambda + rc_0 \in \Lambda$ and $x \in V$, we have

$$\begin{aligned} wf(x) &= f(w^{-1}x) = f(v^{-1}x - \lambda') \\ &= <\lambda, v^{-1}x - \lambda'> + e^{-1}r \\ &= <v\lambda, x> + e^{-1}r - <\lambda, \lambda'> \end{aligned}$$

so that

$$wf = v\lambda + (r - e <\lambda, \lambda'>)c_0$$

is in Λ, since $e<\lambda, \lambda'> \in \mathbb{Z}$ by (1.4.5). □

1.5 Labels

Let S be an irreducible affine root system as in §1.4 and let $W = W(R, L')$. A W-*labelling* k of S is a mapping $k : S \to \mathbb{R}$ such that $k(a) = k(b)$ if a, b are in the same W-orbit in S.

If $S = S(R)$ where R is simply-laced (types A, D, E), all the labels $k(a)$ are equal. If $S = S(R)$ or $S(R)^\vee$ where $R \neq R^\vee$, there are at most two labels, one for short roots and one for long roots. Finally, if S is of type $(C_n^\vee, C_n)$ as in (1.4.3), there are five W-orbits $O_1, \dots, O_5$ in S, as observed in §1.3, and correspondingly five labels $k_1, \dots, k_5$, where $k_i = k(a)$ for $a \in O_i$.

Given a labelling k of S as above, we define a *dual labelling* k' of S', as follows:

(a) if $S = S(R)$, $S' = S(R^\vee)$ (1.4.1) and $a' = \alpha^\vee + rc \in S'$, then $k'(a') = k(\alpha + rc)$.
(b) If $S = S' = S(R)^\vee$ (1.4.2), then $k' = k$.
(c) If $S = S'$ is of type $(C_n^\vee, C_n)$ (1.4.3), the dual labels k_i' ($1 \leq i \leq 5$) are defined by

(1.5.1)
$$\begin{aligned} k_1' &= \tfrac{1}{2}(k_1 + k_2 + k_3 + k_4), \\ k_2' &= \tfrac{1}{2}(k_1 + k_2 - k_3 - k_4), \\ k_3' &= \tfrac{1}{2}(k_1 - k_2 + k_3 - k_4), \\ k_4' &= \tfrac{1}{2}(k_1 - k_2 - k_3 + k_4), \\ k_5' &= k_5, \end{aligned}$$

and $k'(a) = k_i'$ if $a \in O_i$.

In all cases let

$$(1.5.2)\qquad \begin{aligned} \rho_{k'} &= \frac{1}{2}\sum_{\alpha\in R^+} k'(\alpha^\vee)\alpha, \\ \rho'_k &= \frac{1}{2}\sum_{\alpha\in R^+} k(\alpha'^\vee)\alpha'. \end{aligned}$$

where R^+ is the set of positive roots of R determined by the basis (α_i). Explicitly, when $S = S(R)$ (1.4.1) we have

$$\begin{aligned} \rho_{k'} &= \frac{1}{2}\sum_{\alpha\in R^+} k(\alpha)\alpha, \\ \rho'_k &= \frac{1}{2}\sum_{\alpha\in R^+} k(\alpha)\alpha^\vee; \end{aligned}$$

when $S = S(R)^\vee$ (1.4.2) we have

$$\rho_{k'} = \rho'_k = \frac{1}{2}\sum_{\alpha\in R^+} k(\alpha^\vee)\alpha;$$

and when S is of type $(C_n^\vee, C_n)$ (1.4.3), so that R is of type C_n,

$$\begin{aligned} \rho_{k'} &= \sum_{i=1}^n (k'_1 + (n-i)k_5)\varepsilon_i, \\ \rho'_k &= \sum_{i=1}^n (k_1 + (n-i)k_5)\varepsilon_i. \end{aligned}$$

For each $w \in W_0$, we have

$$(1.5.3)\qquad \begin{aligned} w^{-1}\rho'_k &= \frac{1}{2}\sum_{\alpha\in R^+} \sigma(w\alpha)k(\alpha'^\vee)\alpha', \\ w^{-1}\rho_{k'} &= \frac{1}{2}\sum_{\alpha\in R^+} \sigma(w\alpha)k'(\alpha^\vee)\alpha, \end{aligned}$$

where $\sigma(w\alpha) = +1$ or -1 according as $w\alpha \in R^+$ or R^-. In particular, if $i \in I$, $i \neq 0$ we have

$$(1.5.4)\qquad \begin{aligned} s_i\rho'_k &= \rho'_k - k(\alpha_i'^\vee)\alpha'_i, \\ s_i\rho_{k'} &= \rho_{k'} - k'(\alpha_i^\vee)\alpha_i. \end{aligned}$$

(1.5.5) *If the labels $k(\alpha_i'^\vee)$, $k'(\alpha_i^\vee)$ are all nonzero, then ρ'_k and $\rho_{k'}$ are fixed only by the identity element of W_0.* □

Notes and references

Affine root systems were introduced in [M2], which contains an account of their basic properties and their classification. The list of Dynkin diagrams in §1.3 will also be found in the article of Bruhat and Tits [B3] (except that both [M2] and [B3] omit the diagram (1.3.17) when $n = 2$). The reduced affine root systems (1.3.1)–(1.3.14) are in one-one correspondence with the irreducible affine (or Euclidean) Kac-Moody Lie algebras, and correspondingly their diagrams appear in Moody's paper [M9] and Kac's book [K1].

2

The extended affine Weyl group

2.1 Definition and basic properties

Let S be an irreducible affine root system, and let $(a_i)_{i\in I}$ be a basis of S, as in §1.2. For each $i \in I$ let $s_i = s_{a_i}$ be the reflection in the hyperplane H_{a_i} on which a_i vanishes. These reflections generate the Weyl group W_S of S, subject to the relations $s_i{}^2 = 1$ and

$$(s_i s_j)^{m_{ij}} = 1$$

for $i, j \in I$ such that $i \neq j$, whenever $s_i s_j$ has finite order m_{ij}. In other words, W_S is a Coxeter group on the generators s_i, $i \in I$ [B1].

Since W_S rather than S is the present object of study, we may assume that S is reduced, and indeed that $S = S(R)$ where R is a reduced irreducible finite root system spanning a real vector space V of dimension n, as in (1.2.1). We shall say that W_S is of *type* R.

Let $<u, v>$ be a positive-definite symmetric scalar product on V, invariant under the Weyl group of R. We shall regard each root $\alpha \in R$ as a linear function on V, by the rule $\alpha(x) = <\alpha, x>$ for $x \in V$. Then the elements of S are the affine-linear functions $a = \alpha + rc$, where $\alpha \in R$ and $r \in \mathbb{Z}$, and c is the constant function equal to 1.

Let $(\alpha_i)_{i\in I_0}$ be a basis (or set of simple roots) of R, and let R^+ (resp. R^-) denote the set of positive (resp. negative) roots of R determined by this basis. Each $\alpha \in R^+$ is a linear combination of the α_i with non-negative integer coefficients, and there is a unique $\varphi \in R^+$ (the highest root), say

$$\varphi = \sum_{i\in I_0} m_i \alpha_i$$

for which the sum of the coefficients attains its maximum value. The affine roots $a_i = \alpha_i (i \in I_0)$ and $a_0 = -\varphi + c$ then form a basis of $S = S(R)$, and we have

(2.1.1)
$$\sum_{i \in I} m_i a_i = c$$

in conformity with (1.2.2), where $I = I_0 \cup \{0\}$, and $m_0 = 1$.

The alcove C consists of the points $x \in V$ such that $<x, \alpha_i> > 0$ for all $i \neq 0$, and $<x, \varphi> < 1$. It follows that

(2.1.2)
$$S^+ = \{\alpha + rc : \alpha \in R, r \geq \chi(\alpha)\}$$

where χ is the characteristic function of R^-, i.e.,

(2.1.3)
$$\chi(\alpha) = \begin{cases} 0 & \text{if } \alpha \in R^+, \\ 1 & \text{if } \alpha \in R^-. \end{cases}$$

For any $\lambda \in V$, let $t(\lambda) : x \mapsto x + \lambda$ denote translation by λ. In particular, if $\alpha \in R$ we have

$$t(\alpha^\vee) = s_\alpha \cdot s_{\alpha+c} \in W_S$$

where $\alpha^\vee = 2\alpha/|\alpha|^2$. It follows that W_S contains a group of translations isomorphic to the lattice $Q^\vee$ spanned by $R^\vee$, and in fact W_S is the semi direct product

$$W_S = W_0 \ltimes t(Q^\vee)$$

where W_0 is the Weyl group of R, and is the subgroup of W_S that fixes the origin in V.

As in §1.4, let $P^\vee$ be the weight lattice of $R^\vee$ and let L' be either $P^\vee$ as in (1.4.1) and (1.4.2), or $Q^\vee$ if R is of type C_n, as in (1.4.3). The group

(2.1.4)
$$W = W(R, L') = W_0 \ltimes t(L')$$

is called the *extended affine Weyl group*. It coincides with W_S when R is of type E_8, F_4 or G_2, and in the situation of (1.4.3); in all other cases W is larger than W_S. It contains W_S as a normal subgroup, and the quotient $W/W_S \cong L'/Q^\vee = \Omega'$ (1.4.4) is a finite abelian group.

Dually we may define

(2.1.4′)
$$W' = W(R', L) = W_0 \ltimes t(L)$$

and everything in this chapter relating to W applies equally to W'.

Each $w \in W$ is of the form $w = vt(\lambda')$, where $\lambda' \in L'$ and $v \in W_0$. If $a = \alpha + rc \in S$ we have

$$\begin{aligned}(wa)(x) &= a(w^{-1}x) = a(v^{-1}x - \lambda') = <\alpha, v^{-1}x - \lambda'> + r \\ &= <v\alpha, x> + r - <\lambda', \alpha>\end{aligned}$$

for $x \in V$, so that

(2.1.5) $$vt(\lambda')(a) = v(a) - <\lambda', \alpha>c$$

which lies in S because $<\lambda', \alpha> \in \mathbb{Z}$. It follows that W permutes S.

For each $i \in I$, $i \neq 0$, let

$$<L', \alpha_i> = \{<\lambda', \alpha_i> : \lambda' \in L'\},$$

a subgroup of $\mathbb{Z}$. Since $\alpha_i^\vee \in L'$ it follows that $2 \in <L', \alpha_i>$, and hence that $<L', \alpha_i> = \mathbb{Z}$ or $2\mathbb{Z}$. If $<L', \alpha_i> = 2\mathbb{Z}$ then $<\alpha_j^\vee, \alpha_i>$ is an even integer for all $j \neq 0$, and a consideration of Dynkin diagrams shows that

(2.1.6) *We have* $<L', \alpha_i> = 2\mathbb{Z}$ *only in the following situation: R is of type C_n, $L' = Q^\vee$, and α_i is the unique long simple root of R* (i.e., $\alpha_i = 2\varepsilon_n$ *in the notation of* (1.3.4)). *In all other cases,* $<L', \alpha_i> = \mathbb{Z}$.

2.2 The length function on W

For each $w \in W$ let

(2.2.1) $$S(w) = S^+ \cap w^{-1}S^-$$

so that $a \in S(w)$ if and only if $a(x) > 0$ and $a(w^{-1}x) < 0$ for $x \in C$, that is to say if and only if the hyperplane H_a separates the alcoves C and $w^{-1}C$. It follows that $S(w)$ is a finite set, and we define the *length* of $w \in W$ to be

$$l(w) = \text{Card } S(w).$$

From (2.2.1) it follows that

(2.2.2) $$S(w^{-1}) = -wS(w)$$

and hence that $l(w^{-1}) = l(w)$.

In particular, we have

(2.2.3) $$S(s_i) = \{a_i\}$$

for all $i \in I$, and hence $l(s_i) = 1$.

Since W permutes S, it permutes the hyperplanes $H_a (a \in S)$ and hence also the alcoves. Hence for each $w \in W$ there is a unique $v \in W_S$ such that $wc = vc$, and therefore $u = v^{-1}w$ stabilizes C and so permutes the a_i $(i \in I)$. We have $l(w) = l(v)$ and $l(u) = 0$.

Let

$$\Omega = \{u \in W : l(u) = 0\}.$$

From above it follows that $W = W_S \rtimes \Omega$, so that (1.4.4)

$$\Omega \cong W/W_S \cong L'/Q^\vee = \Omega'$$

is a finite obelian group. Later (§2.5) we shall determine the elements of Ω explicitly.

(2.2.4) *Let $v, w \in W$. Then*

$$l(vw) \le l(v) + l(w)$$

and the following conditions are equivalent:

(i) $l(v) + l(w) = l(vw)$,
(ii) $S(vw) = w^{-1}S(v) \cup S(w)$,
(iii) $w^{-1}S(v) \subset S^+$,
(iv) $S(w) \subset S(vw)$,
(v) $w^{-1}S(v) \subset S(vw)$.

Proof Let

$$X = S^+ \cap w^{-1}S^+ \cap w^{-1}v^{-1}S^-,$$
$$Y = S^+ \cap w^{-1}S^- \cap w^{-1}v^{-1}S^+,$$
$$Z = S^+ \cap w^{-1}S^- \cap w^{-1}v^{-1}S^-.$$

The four sets X, Y, $-Y$ and Z are pairwise disjoint. (For example, X is contained in $w^{-1}S^+$ and Y in $w^{-1}S^-$.) We have

$$w^{-1}S(v) = X \cup -Y, \quad S(w) = Y \cup Z, \quad S(vw) = X \cup Z.$$

Hence

$$l(v) + l(w) - l(vw) = 2 \text{ Card } Y \ge 0$$

and each of the conditions (i)–(v) is equivalent to $Y = \emptyset$. □

From (2.2.4) it follows in particular that

(2.2.5) $$S(uw) = S(w), \quad S(wu) = u^{-1}S(w)$$

if $w \in W$ and $u \in \Omega$.

(2.2.6) *Let $v, w \in W$. Then $S(v) = S(w)$ if and only if $vw^{-1} \in \Omega$.*

Proof If $vw^{-1} \in \Omega$, (2.2.5) shows that $S(v) = S(w)$. Conversely, replacing w by w^{-1}, we have to show that $S(v) = S(w^{-1})$ implies $vw \in \Omega$. From the proof of (2.2.4) we have

$$X \cup -Y = w^{-1}S(w^{-1}) = -S(w) = -Y \cup -Z$$

so that $X = -Z$ and therefore $X = Z = \emptyset$, since both X and Z are subsets of S^+. Hence $S(vw) = \emptyset$, i.e., $vw \in \Omega$. □

For $a \in S$, let

(2.2.7) $$\sigma(a) = \begin{cases} +1 & \text{if } a \in S^+, \\ -1 & \text{if } a \in S^-. \end{cases}$$

(2.2.8) *Let $w \in W, i \in I$. Then*

(i) $l(s_i w) = l(w) + \sigma(w^{-1}a_i)$,
(ii) $l(ws_i) = l(w) + \sigma(wa_i)$.

Proof (i) From (2.2.4) with $v = s_i$ we have $l(s_i w) = l(w) + 1$ if and only if $w^{-1}S(s_i) \subset S^+$, i.e. if and only if $\sigma(w^{-1}a_i) = 1$. By replacing w by $s_i w$, it follows that $l(w) = l(s_i w)$ if and only if $\sigma(w^{-1}a_i) = -1$.
(ii) Since $l(ws_i) = l(s_i w^{-1})$ and $l(w) = l(w^{-1})$, this follows from (i). □

Let $l(w) = p > 0$. Then $w \notin \Omega$, hence $wa_i \in S^-$ for some $i \in I$. By (2.2.8) we have $l(ws_i) = p - 1$. By induction on p it follows that w may be written in the form

$$w = us_{i_1} \cdots s_{i_p}$$

where $i_1, \ldots, i_p \in I$ and $u \in \Omega$. Such an expression (with $p = l(w)$) is called a *reduced expression* for w.

(2.2.9) *For w as above,*

$$S(w) = \{b_1, \ldots, b_p\}$$

where

$$b_r = s_{i_p} \cdots s_{i_{r+1}}(a_{i_r}) \qquad (1 \le r \le p).$$

Proof If $p = 0$, then $w \in \Omega$ and $S(w)$ is empty. If $p \geq 1$ let $v = ws_{i_p}$, then as above $l(v) = p - 1$ and therefore

$$S(w) = s_{i_p} S(v) \cup \{a_{i_p}\}$$

by (2.2.4). Hence the result follows by induction on p. □

2.3 The Bruhat order on W

Since W_S is a Coxeter group it possesses a Bruhat ordering, denoted by $v \leq w$ (see e.g. [B1] ch.5) We extend this partial ordering to the extended affine Weyl group W as follows. If $w = uv$ and $w' = u'v'$ are elements of W, where $u, u' \in \Omega$ and $v', v' \in W_S$, then we define

(2.3.1) $w \leq w'$ *if and only if* $u = u'$ *and* $v \leq v'$.

Thus the distinct cosets of W_S in W are incomparable for this ordering.

From standard properties of the Bruhat ordering on a Coxeter group (loc. cit.) it follows that

(2.3.2) *Let* $v, w \in W$ *and let* $w = us_{i_1} \cdots s_{i_p}$ *be a reduced expression for* w (*so that* $u \in \Omega$ *and* $p = l(w)$). *Then the following conditions are equivalent:*

(a) $v \leq w$;
(b) *there exists a subsequence* $(j_1, \ldots, j_q)$ *of the sequence* $(i_1, \ldots, i_p)$ *such that* $v = us_{j_1} \cdots s_{j_q}$;
(c) *there exists a subsequence* $(j_1, \ldots, j_q)$ *of the sequence* $(i_1, \ldots, i_p)$ *such that* $v = us_{j_1} \cdots s_{j_q}$ *is a reduced expression for* v.

(2.3.3) *Let* $w \in W, a \in S^+$. *Then the following are equivalent:*
(a) $a \in S(w)$; (b) $l(ws_a) < l(w)$; (c) $ws_a < w$.

Proof Let $w = us_{i_1} \cdots s_{i_p}$ be a reduced expression. If $wa \in S^-$, then $a = b_r$ for some r, in the notation of (2.2.9), so that $s_a = s_{i_p} \cdots s_{i_{r+1}} s_{i_r} s_{i_{r+1}} \cdots s_{i_p}$, and therefore

$$ws_a = us_{i_1} \cdots s_{i_{r-1}} s_{i_{r+1}} \cdots s_{i_p} < w.$$

It follows that (a) $\Rightarrow$ (c) $\Rightarrow$ (b). Conversely, if $wa \in S^+$ then $(ws_a)a = -wa \in S^-$, and hence $l(w) < l(ws_a)$ by the previous argument applied to ws_a. Hence (b) $\Rightarrow$ (a). □

2.4 The elements $u(\lambda')$, $v(\lambda')$

We shall first compute the length of an arbitrary element of W. As before (2.1.3), let χ be the characteristic function of R^-.

(2.4.1) *Let* $\lambda' \in L'$, $w \in W_0$. *Then*

(i) $l(wt(\lambda')) = \sum_{\alpha \in R^+} |<\lambda', \alpha> + \chi(w\alpha)|$,

(ii) $l(t(\lambda')w) = \sum_{\alpha \in R^+} |<\lambda', \alpha> - \chi(w^{-1}\alpha)|$,

Proof (i) Let $a = \alpha + rc \in S$. From (2.1.5) we have

$$wt(\lambda')(a) = w\alpha + (r - <\lambda', \alpha>)c,$$

so that by (2.1.2) $a \in S(wt(\lambda'))$ if and only if

$$\chi(\alpha) \leq r < \chi(w\alpha) + <\lambda', \alpha>. \tag{1}$$

For each $\alpha \in R$, let

$$f(\alpha) = <\lambda', \alpha> + \chi(w\alpha) - \chi(\alpha).$$

Then it follows from (1) that the number of roots $a \in S(wt(\lambda'))$ with gradient α is equal to $f(\alpha)$ if $f(\alpha) \geq 0$, and is zero otherwise. Since $f(\alpha) + f(-\alpha) = 0$, we have

$$\begin{aligned} l(wt(\lambda')) &= \frac{1}{2}\sum_{\alpha \in R} |f(\alpha)| = \sum_{\alpha \in R^+} |f(\alpha)| \\ &= \sum_{\alpha \in R^+} |<\lambda', \alpha> + \chi(w\alpha)|. \end{aligned}$$

(ii) This follows from (i), since $l(t(\lambda')w) = l(w^{-1}t(-\lambda'))$ by (2.2.2). □

For each $\lambda' \in L'$, let λ'_+ denote the unique dominant weight in the orbit $W_0\lambda'$. Then it follows from (2.4.1) that

$$l(t(\lambda')) = l(t(\lambda'_+)) = \sum_{\alpha \in R^+} <\lambda'_+, \alpha>. \tag{2.4.2}$$

Let w_0 be the longest element of W_0, and let $\lambda'_- = w_0\lambda'_+$ be the antidominant weight in the orbit $W_0\lambda'$. Let $v(\lambda')$ be the shortest element of W_0 such that $v(\lambda')\lambda' = \lambda'_-$, and define $u(\lambda') \in W$ by $u(\lambda') = t(\lambda')v(\lambda')^{-1}$. Thus we have

$$t(\lambda') = u(\lambda')v(\lambda'), \quad t(\lambda'_-) = v(\lambda')t(\lambda')v(\lambda')^{-1} = v(\lambda')u(\lambda'). \tag{2.4.3}$$

We have

(2.4.4) $$S(v(\lambda')) = \{\alpha \in R^+ : <\lambda', \alpha> > 0\}.$$

Proof Let $v(\lambda') = s_{i_1} \cdots s_{i_p}$ be a reduced expression. By (2.2.9), $S(v(\lambda')) = \{\beta_1, \dots, \beta_p\}$, where $\beta_r = s_{i_p} \cdots s_{i_{r+1}}(\alpha_{i_r})$ for $1 \le r \le p$. Let

$$\lambda'_r = s_{i_{r+1}} \cdots s_{i_p} \lambda' = s_{i_r} \cdots s_{i_1} \lambda'_-$$

for $0 \le r \le p$, so that $\lambda'_0 = \lambda'_-$ and $\lambda'_p = \lambda'$.

Suppose that $\lambda'_{r-1} = \lambda'_r$ for some r. Then

$$\lambda'_- = s_{i_1} \cdots s_{i_{r-1}} \lambda'_{r-1} = s_{i_1} \cdots s_{i_{r-1}} \lambda'_r = w\lambda'$$

where $w = s_{i_1} \cdots s_{i_{r-1}} s_{i_{r+1}} \cdots s_{i_p}$ is shorter than $v(\lambda')$. It follows that $\lambda'_r \ne \lambda'_{r-1} = s_{i_r} \lambda'_r$, so that $<\lambda'_r, \alpha_{i_r}> \ne 0$. But

$$<\lambda'_r, \alpha_{i_r}> = <\lambda', s_{i_p} \cdots s_{i_{r+1}} \alpha_{i_r}> = <\lambda', \beta_r>$$

and $<\lambda', \beta_r> = <\lambda'_-, v(\lambda')\beta_r>$ is ≥ 0, because $v(\lambda')\beta_r \in R^-$. Hence $<\lambda', \beta_r> > 0$ for $1 \le r \le p$.

Conversely, if $\beta \in R^+$ and $<\lambda', \beta> > 0$, we have $<\lambda'_-, v(\lambda')\beta> > 0$ and therefore $v(\lambda')\beta \in R^-$, i.e., $\beta \in s(v(\lambda'))$. □

(2.4.5) *$u(\lambda')$ is the shortest element of the coset $t(\lambda')W_0$, and*

$$l(t(\lambda')) = l(u(\lambda')) + l(v(\lambda'))$$

for all $\lambda' \in L'$.

Proof It follows from (2.4.1) (ii) that, for fixed $\lambda' \in L'$ and varying $w \in W_0$, the length of $t(\lambda')w^{-1}$ will be least if, for each $\alpha \in R^+$, $\chi(w\alpha) = 1$ if and only if $<\lambda', \alpha> > 0$, i.e. (by (2.4.4)) if and only if $S(w) = S(v(\lambda'))$. By (2.2.6), this forces $w = v(\lambda')$ and proves the first statement. It now follows from (2.4.1) (ii) and (2.4.4) that

$$\begin{aligned} l(u(\lambda')) = l(t(\lambda')v(\lambda')^{-1}) &= \sum_{\alpha \in R^+} |<\lambda', \alpha> - \chi(v(\lambda')\alpha)| \\ &= l(t(\lambda')) - l(v(\lambda')). \end{aligned}$$ □

From (2.4.4) it follows that, for all $\alpha \in R$

(2.4.6) $\chi(v(\lambda')\alpha) = 1$ *if and only if* $\chi(\alpha) + <\lambda', \alpha> > 0$. □

(2.4.7) *Let* $\alpha \in R$, $\beta = v(\lambda')^{-1}\alpha$, $r \in \mathbb{Z}$. *Then*

(i) $\alpha + rc \in S(u(\lambda'))$ *if and only if* $\alpha \in R^-$ *and* $1 \leq r < \chi(\beta) + <\lambda', \beta>$.
(ii) $\alpha + rc \in S(u(\lambda')^{-1})$ *if and only if* $\chi(\alpha) \leq r < -<\lambda', \alpha>$.

Proof (i) Since $u(\lambda') = v(\lambda')^{-1}t(\lambda'_-)$ (2.4.3), it follows from (2.1.5) that $\alpha + rc \in S(u(\lambda'))$ if and only if

$$\chi(\alpha) \leq r < \chi(\beta) + <\lambda', \beta>,$$

since $<\lambda'_-, \alpha> = <\lambda', \beta>$. Hence $\chi(\beta) + <\lambda', \beta> > 0$ and therefore $\chi(\alpha) = 1$ by (2.4.6).
(ii) We have $u(\lambda')^{-1} = v(\lambda')t(-\lambda')$, hence $\alpha + rc \in S(u(\lambda')^{-1})$ if and only if

$$\chi(\alpha) \leq r < \chi(v(\lambda')\alpha) - <\lambda', \alpha>.$$

Hence $\chi(\alpha) + <\lambda', \alpha> \leq 0$ and therefore $\chi(v(\lambda')\alpha) = 0$ by (2.4.6). □

(2.4.8) *Let* $a \in S^+$. *Then* $a \in S(u(\lambda')^{-1})$ *if and only if* $a(\lambda') < 0$.

This is a restatement of (2.4.7) (ii).

(2.4.9) *Let* $w \in W_0$, $\lambda' \in L'$. *Then*

$$l(u(\lambda')w) = l(u(\lambda')) + l(w).$$

Proof By (2.2.4) it is enough to show that $w^{-1}S(u(\lambda')) \subset S^+$, and this follows from (2.4.7) (i). □

(2.4.10) *Let* $w \in W$, $w(0) = \lambda'$. *Then* $w \geq u(\lambda')$.

Proof We have $u(\lambda')(0) = \lambda'$, hence $w \geq u(\lambda')v$ for some $v \in W_0$. Now apply (2.4.9). □

(2.4.11) *Let* φ *be the highest root of* R. *Then* $u(\varphi^\vee) = s_0$, *and* $v(\varphi^\vee) = s_\varphi$.

Proof We have $t(\varphi^\vee) = s_0 s_\varphi$, and $l(s_0) = 1$. Hence s_0 is the shortest element of the coset $t(\varphi^\vee)W_0$, hence is equal to $u(\varphi^\vee)$ by (2.4.5). It then follows that $v(\varphi^\vee) = s_\varphi$. □

(2.4.12) *Let* $\lambda' \in L'$, $v \in \Omega$, $\mu' = v\lambda'$. *Then* $u(\mu') = vu(\lambda')$.

Proof Since $vu(\lambda')(0) = \mu'$ we have $u(\mu') \leq vu(\lambda')$ by (2.4.10), hence $l(u(\mu')) \leq l(u(\lambda'))$ Replacing (λ', v) by (μ', v^{-1}) gives the reverse inequality $l(u(\lambda')) \leq l(u(\mu'))$. Hence $u(\mu') = vu(\lambda')$. □

(2.4.13) *Let $w \in W$. Then $S(w) \cap R = \emptyset$ if and only if $w = u(\lambda')$, where $\lambda' = w(0)$.*

Proof We have $w = u(\lambda')v$ with $v \in w_0$. From (2.4.9) and (2.2.4) it follows that $S(w) = v^{-1}S(u(\lambda')) \cup S(v)$. By (2.4.7) (i), $S(u(\lambda')) \cap R = \emptyset$; hence $S(w) \cap R = \emptyset$ if and only if $S(v) = \emptyset$, i.e., $v = 1$. □

(2.4.14) *Let $\lambda' \in L'$, $i \in I$, $\mu' = s_i\lambda'$.*
(i) *If $a_i(\lambda') \neq 0$, then $u(\mu') = s_iu(\lambda')$ and $v(\mu') = v(\lambda')s_{\alpha_i}$.*
(ii) *If $a_i(\lambda') < 0$, then $u(\lambda') > u(\mu')$, and $v(\lambda') < v(\mu')$* if $i \neq 0$.
(iii) *If $a_i(\lambda') = 0$, then $s_iu(\lambda') = u(\lambda')s_j$ for some $j \neq 0$, and $v(\lambda')\alpha_i = \alpha_j$.*

Proof (i) By interchanging λ' and μ' if necessary, we may assume that $a_i(\lambda') < 0$, so that $u(\lambda')^{-1}a_i \in S^-$ by (2.4.8). If $w = s_iu(\lambda')$ we have $l(w) = l(u(\lambda'))-1$, hence $l(u(\lambda')^{-1}) = l(w^{-1}) + l(s_i)$ and therefore $s_iS(w^{-1}) \subset S(u(\lambda')^{-1})$ by (2.2.4) (v). It follows that for each $b \in S(w^{-1})$ we have $(s_ib)(\lambda') < 0$ by (2.4.8), that is to say $b(\mu') < 0$ and therefore $b \in S(u(\mu')^{-1})$. Hence $S(w^{-1}) \subset S(u(\mu')^{-1})$ and therefore $l(w) \leq l(u(\mu'))$. But $w(0) - s_i\lambda' - \mu'$, so that $w = u(\mu')$ by (2.4.5), i.e. $u(\mu') = s_iu(\lambda')$. Consequently

$$\begin{aligned} v(\mu') = u(\mu')^{-1}t(\mu') &= u(\mu')^{-1}s_it(\lambda')s_{\alpha_i} \\ &= u(\lambda')^{-1}t(\lambda')s_{\alpha_i} = v(\lambda')s_{\alpha_i}. \end{aligned}$$

(ii) From above we have $u(\lambda') = s_iu(\mu')$ and $l(u(\lambda')) = l(u(\mu')) + 1$, so that $u(\lambda') > u(\mu')$. If $i \neq 0, l(t(\mu')) = l(t(\lambda'))$ by (2.4.2), so that $l(v(\lambda')) = l(v(\mu')) - 1$ and therefore $v(\lambda') < v(\mu')$.
(iii) If $a_i(\lambda') = 0$ then $s_iu(\lambda')(0) = s_i\lambda' = \lambda'$, and therefore $s_iu(\lambda') = u(\lambda')w$ for some $w \in W_0$. By (2.4.9), $l(s_iu(\lambda')) = l(u(\lambda')) + l(w)$, hence $l(w) = 1$ and therefore $w = s_j$ for some $j \in I, j \neq 0$. It follows that $s_it(\lambda')v(\lambda')^{-1} = t(\lambda')v(\lambda')^{-1}s_j$. Now $s_it(\lambda')s_{\alpha_i} = t(s_i\lambda') = t(\lambda')$, and hence $s_{\alpha_i}v(\lambda')^{-1} = v(\lambda')^{-1}s_j$, so that $v(\lambda')\alpha_i = \pm\alpha_j$. But $v(\lambda')\alpha_i \in R^+$ by (2.4.4), hence $v(\lambda')\alpha_i = \alpha_j$. □

2.5 The group Ω

We shall now determine the elements of the finite group

$$\Omega = \{w \in W : l(w) = 0\}.$$

For this purpose let $\pi_i'(i \in I_0)$ be the fundamental weights of $R^\vee$, defined by the relations

(2.5.1) $$<\pi_i', \alpha_j> = \delta_{ij}$$

for $i, j \in I_0$; also, to complete the notation, let $\pi_0' = 0$. We define

(2.5.2) $$u_i = u(\pi_i'),\ v_i = v(\pi_i')$$

(so that in particular $u_0 = v_0 = 1$). Next, let J be the subset of I defined by

(2.5.3) $j \in J$ *if and only if* $\pi_j' \in L'$ *and* $m_j = 1$,

where the positive integers m_j are those defined in (2.1.1), so that $m_0 = 1$ and $m_j = <\pi_j', \varphi>$ for $j \neq 0$, where φ is the highest root of R. We have $0 \in J$ in all cases.

With this notation established, we have

(2.5.4) $$\Omega = \{u_j : j \in J\}.$$

Proof Let $u \in \Omega$. Clearly u is the shortest element of its coset uW_0, so that $u = u(\lambda')$ where $\lambda' = u(0)$. Hence $u = t(\lambda')v(\lambda')^{-1}$ and it follows from (2.4.1) (ii) that $<\lambda', \alpha> = \chi(v(\lambda')\alpha)$ for each $\alpha \in R^+$. Hence λ' is dominant and of the form $\lambda' = \sum_{i\neq 0} c_i\pi_i'$, where the coefficients c_i are integers ≥ 0. Hence

(1) $$<\lambda', \varphi> = \sum_{i\neq 0} c_i m_i.$$

On the other hand, $<\lambda', \varphi> = \chi(v(\lambda')\varphi) = 0$ or 1. If $<\lambda', \varphi> = 0$ it follows from (1) that each $c_i = 0$, hence $\lambda' = 0$ and $u = 1$; if $<\lambda', \varphi> = 1$ it follows that $\lambda' = \pi_j'$ for some $j \neq 0$ such that $m_j = 1$. Hence $u = u_j$ for some $j \in J$.

Conversely, let us show that $u_j \in \Omega$ for each $j \in J$. Each root $\alpha \in R^+$ is of the form

$$\alpha = \sum_{i\neq 0} m_i' \alpha_i$$

where $0 \leq m_i' \leq m_i$. Hence $0 \leq <\pi_i', \alpha> \leq <\pi_i', \varphi>$. In particular, if $j \in J$ (and $j \neq 0$) we have $<\pi_j', \alpha> = 0$ or 1 for each $\alpha \in R^+$, and from (2.4.6) it follows that $<\pi'_j, \alpha> = \chi(v_j\alpha)$ for each $\alpha \in R^+$. Hence by (2.4.1) (ii)

$$l(u_j) = l\big(t(\pi'_j)v_j^{-1}\big) = \sum_{\alpha\in R^+} |<\pi'_j, \alpha> - \chi(v_j\alpha)| = 0$$

and therefore $u_j \in \Omega$. □

(2.5.5) *Let $j \in J$. Then $u_j(a_0) = a_j$.*

Proof Since u_j has length zero, it permutes the simple affine roots a_i. Hence $u_j(a_0) = a_r$ for some $r \in I$, and therefore

(1) $$v_j^{-1}a_0 = t(\pi'_j)^{-1}a_r = a_r + <\pi'_j, \alpha_r>c$$

(where $\alpha_0 = -\varphi$ if $r = 0$).

Evaluating both sides of (1) at the origin gives

$$a_r(0) + <\pi'_j, \alpha_r> = 1.$$

If $r \neq 0$ we have $<\pi'_j, \alpha_r> = 1$ and hence $r = j$. If $r = 0$ we obtain $<\pi'_j, \alpha_0> = 0$, hence $j = 0$. □

For $j, k \in J$ we define $j + k$ and $-j$ by requiring that

(2.5.6) $$u_{j+k} = u_j u_k, \quad u_{-j} = u_j^{-1}$$

thereby making J an abelian group with neutral element 0, isomorphic to Ω. Likewise, for $i \in I$ and $j \in J$ we define $i + j \in I$ by requiring that

(2.5.7) $$u_j(a_i) = a_{i+j}.$$

(If $i \in J$, the two definitions agree, by virtue of (2.5.5).) Thus J acts on I as a group of permutations.

(2.5.8) *Let $i \in I$, $j \in J$. Then $v_j\alpha_i = \alpha_{i-j}$.*

Proof We have

$$u_j v_j a_i = t(\pi'_j)a_i = a_i - <\alpha_i, \pi'_j>c$$

by (2.1.5). Hence

$$\begin{aligned} v_j a_i &= u_{-j}a_i - <\alpha_i, \pi'_j>c \\ &= a_{i-j} - <\alpha_i, \pi'_j>c, \end{aligned}$$

so that $D(v_j a_i) = D(a_{i-j})$, i.e. $v_j\alpha_i = \alpha_{i-j}$. □

From (2.5.6) it follows that if $j, k \in J$

$$\begin{aligned} t(\pi'_{j+k})v_{j+k}^{-1} &= t(\pi'_j)v_j^{-1}t(\pi'_k)v_k^{-1} \\ &= t(\pi'_j + v_j^{-1}\pi'_k)v_j^{-1}v_k^{-1} \end{aligned}$$

and therefore

(2.5.9) $$\pi'_{j+k} = \pi'_j + v_j^{-1}\pi'_k = \pi'_k + v_k^{-1}\pi'_j,$$

(2.5.10) $$v_{j+k} = v_j v_k = v_k v_j, \; v_{-j} = v_j^{-1}.$$

More generally, if $i \in I$ and $j \in J$ we have

(2.5.11) $$\pi'_{i+j} = m_i \pi'_j + v_j^{-1}\pi'_i.$$

Proof This is clear if $i = 0$ or $j = 0$, so we may assume that $i \neq 0$ and $j \neq 0$. Let $k \in I$, $k \neq 0$. From (2.5.8) we have $v_j\alpha_k = \alpha_{k-j}$, so that

$$< v_j^{-1}\pi'_i, \alpha_k > \; = \; < \pi'_i, v_j\alpha_k > \; = \; < \pi'_i, \alpha_{k-j} >$$

is zero unless $k = j$ or $k = i + j$. If $k = i + j$ it is equal to 1, and if $k = j$ it is equal to $< \pi'_i, \alpha_0 > \; = -m_i$. Hence

$$v_j^{-1}\pi'_i = \pi'_{i+j} - m_i\pi'_j.$$ □

Finally, let w_0 be the longest element of W_0, and for each $j \in J$ let w_{0j} be the longest element of the isotropy subgroup W_{0j} of π'_j in W_0. Then we have

(2.5.12) $$v_j = w_0 w_{0j}.$$

For $w_0 w_{0j}$ is the shortest element of W_0 that takes π'_j to $w_0\pi'_j$.

2.6 Convexity

Let

$$Q_+^\vee = \sum_{i \neq 0} \mathbb{N}\alpha_i^\vee$$

denote the cone in $Q^\vee$ spanned by the simple coroots $\alpha_i^\vee$, and let L'_{++} denote the set of dominant weights $\lambda' \in L'$, satisfying $< \lambda', \alpha_i > \; \geq 0$ for $i \neq 0$. As in §2.4, for each $\lambda' \in L'$ let λ'_+ denote the unique dominant weight in the orbit $W_0\lambda'$.

A subset X of L' is said to be *saturated* if for each $\lambda' \in X$ and each $\alpha \in R$ we have $\lambda' - r\alpha^\vee \in X$ for all integers r between 0 and $< \lambda', \alpha >$ (inclusive). In other words, the segment $[\lambda', s_\alpha\lambda'] \cap (\lambda' + Q^\vee)$ is contained in X. In particular, $s_\alpha\lambda' \in X$, so that a saturated set is W_0-stable.

The intersection of any family of saturated sets is saturated. In particular, given any subset of L', there is a smallest saturated set containing it.

(2.6.1) *Let X be a saturated subset of L', and let $\lambda' \in X$. If $\mu' \in L'_{++}$ is such that $\lambda' - \mu' \in Q^\vee_+$, then $\mu' \in X$.*

Proof Let $\nu' = \lambda' - \mu' = \sum r_i\alpha_i^\vee$. We proceed by induction on $r = r(\nu') = \sum r_i$. If $r = 0$, then $\lambda' = \mu'$ and there is nothing to prove. Let $r \geq 1$, then $\nu' \neq 0$ and therefore $\sum r_i <\nu', \alpha_i^\vee> = |\nu'|^2 > 0$. Hence for some $i \neq 0$ we have $r_i \geq 1$ and $<\nu', \alpha_i> \geq 1$. Since $<\mu', \alpha_i> \geq 0$ it follows that $<\lambda', \alpha_i> \geq 1$ and hence that $\lambda'_1 = \lambda' - \alpha_i^\vee \in X$. Consequently $\mu' = \lambda'_1 - \nu'_1$, where $\nu'_1 = \nu' - \alpha_i^\vee \in Q'_+$ and $r(\nu'_1) = r - 1$. By the inductive hypothesis it follows that $\mu' \in X$. □

Let $\lambda' \in L'_{++}$ and let $\Sigma(\lambda')$ denote the smallest saturated subset of L' that contains λ'. Let $C(\lambda')$ denote the convex hull in V of the orbit $W_0\lambda'$, and let

$$\Sigma_1(\lambda') = C(\lambda') \cap (\lambda' + Q^\vee),$$
$$\Sigma_2(\lambda') = \bigcap_{w \in W_0} w(\lambda' - Q^\vee_+).$$

Then we have

$$\Sigma(\lambda') = \Sigma_1(\lambda') = \Sigma_2(\lambda'). \tag{2.6.2}$$

Proof (a) $\Sigma(\lambda') \subset \Sigma_1(\lambda')$. Since $\lambda' \in \Sigma_1(\lambda')$, it is enough to show that $\Sigma_1(\lambda')$ is saturated. Now both $C(\lambda')$ and $\lambda' + Q^\vee$ are W_0-stable, hence $\Sigma_1(\lambda')$ is W_0-stable and therefore contains $s_\alpha\lambda'$ for each $\alpha \in R$. By convexity, $\Sigma_1(\lambda')$ contains the interval $[\lambda', s_\alpha\lambda'] \cap (\lambda' + Q^\vee)$, hence is saturated.
(b) $\Sigma_1(\lambda') \subset \Sigma_2(\lambda')$. Each set $w(\lambda' - Q^\vee_+)$ is the intersection of $\lambda' + Q^\vee$ with a convex set, hence the some is true of $\Sigma_2(\lambda')$. Moreover, $\Sigma_2(\lambda')$ contains the orbit $W_0\lambda'$, since $\lambda' - w\lambda' \in Q^\vee_+$ for all $w \in W_0$. Hence $\Sigma_2(\lambda')$ contains $\Sigma_1(\lambda')$.
(c) $\Sigma_2(\lambda') \subset \Sigma(\lambda')$. If $\mu' \in \Sigma_2(\lambda')$, let λ'_+ be the dominant element of the orbit $W_0\mu'$. Then $\mu'_+ \in \lambda' - Q^\vee_+$, hence $\mu'_+ \in \Sigma(\lambda')$ by (2.6.1). Since $\Sigma(\lambda')$ is W_0-stable, it follows that $\mu' \in \Sigma(\lambda')$. Hence $\Sigma_2(\lambda') \subset \Sigma(\lambda')$, and the proof is complete. □

(2.6.3) *Let $\lambda', \mu' \in L'_{++}$. Then the following conditions are equivalent:* (a) $\lambda' - \mu' \in Q^\vee_+$; (b) $\mu' \in \Sigma(\lambda')$; (c) $\Sigma(\mu') \subset \Sigma(\lambda')$.

Proof (2.6.1) shows that (a) implies (b), and it is clear that (b) and (c) are equivalent. Finally, if $\Sigma(\mu') \subset \Sigma(\lambda')$ then $\mu' \in \Sigma(\lambda') = \Sigma_2(\lambda')$, hence $\mu' \in \lambda' - Q_+^\vee$, so that (c) implies (a). □

If $\lambda', \mu' \in L'_{++}$ satisfy the equivalent conditions of (2.6.3) we write

(2.6.4) $$\lambda' \geq \mu'.$$

This is the *dominance partial ordering* on L'_{++}.

2.7 The partial order on L'

Recall (§2.4) that for $\lambda' \in L'$ the shortest $w \in W_0$ such that $w\lambda' = \lambda'_-$ is denoted by $v(\lambda')$. Also let $\bar{v}(\lambda')$ denote the shortest $w \in W_0$ such that $w\lambda'_+ = \lambda'$. Here λ'_+ is the dominant weight and $\lambda'_- = w_0\lambda'_+$ the antidominant weight in the orbit $W_0\lambda'$, and w_0 is the longest element of W_0. We have

(2.7.1) $$v(\lambda')^{-1} = w_0\bar{v}(\mu')w_0 = \bar{v}(-\lambda'), \textit{ where } \mu' = w_0\lambda'.$$

Proof Since λ' and μ' are in the same W_0-orbit we have $\mu'_+ = \lambda'_+$ and $\mu'_- = \lambda'_-$. Hence $w_0\bar{v}(\mu')w_0\lambda'_- = w_0\bar{v}(\mu')\mu'_+ = w_0\mu' = \lambda'$, and $w_0\bar{v}(\mu')w_0$ is the shortest element of W_0 with this property. It follows that $w_0\bar{v}(\mu')w_0 = v(\lambda')^{-1}$.

Next, let $\nu' = -\lambda'$. Then $-\nu'_+ = \lambda'_-$, and again by minimality it follows that $\bar{v}(\nu') = v(\lambda')^{-1}$. □

(2.7.2) (i) $S(v(\lambda')) = \{\alpha \in R^+ : \langle\lambda', \alpha\rangle > 0\}$,
(ii) $S(\bar{v}(\lambda')^{-1}) = \{\alpha \in R^+ : \langle\lambda', \alpha\rangle < 0\}$.

Proof (i) is a restatement of (2.4.4), and (ii) follows from (i) and (2.7.1), since $\bar{v}(\lambda')^{-1} = v(-\lambda')$. □

(2.7.3) *Let* $\lambda' \in L'$. *Then* $v(\lambda')\bar{v}(\lambda') = v(\lambda'_+)$.

Proof Since $v(\lambda')\, \bar{v}(\lambda')$ sends λ'_+ to λ'_-, it follows that $v(\lambda')\bar{v}(\lambda') \geq v(\lambda'_+)$. On the other hand, by (2.7.2) we have

$$\begin{aligned} l(v(\lambda'_+)) &= \text{Card}\,\{\alpha \in R^+ : \langle\lambda'_+, \alpha\rangle > 0\} \\ &= \text{Card}\,\{\alpha \in R^+ : \langle\lambda', \alpha\rangle \neq 0\} \\ &= l(v(\lambda')) + l(\bar{v}(\lambda')) \geq l(v(\lambda')\bar{v}(\lambda')). \end{aligned}$$

Hence $v(\lambda')\bar{v}(\lambda') = v(\lambda'_+)$. □

(2.7.4) *Let $\lambda', \mu' \in L'$ be in the same W_0-orbit. Then $\bar{v}(\lambda') \geq \bar{v}(\mu')$ if and only if $v(\lambda') \leq v(\mu')$.*

Proof Let w be the longest element of W_0 that fixes λ'_+, so that $w = w^{-1}$, and $v(\lambda'_+) = w_0 w$. From (2.7.3) we have $v(\lambda')\bar{v}(\lambda') = v(\mu')\bar{v}(\mu') = w_0 w$. Since $l(\bar{v}(\lambda')w) = l(\bar{v}(\lambda')) + l(w)$, and likewise for μ', we have

$$\begin{aligned} \bar{v}(\lambda') \geq \bar{v}(\mu') &\iff \bar{v}(\lambda')w \geq \bar{v}(\mu')w \\ &\iff v(\lambda')^{-1}w_0 \geq v(\mu')^{-1}w_0 \\ &\iff v(\lambda')^{-1} \leq v(\mu')^{-1} \\ &\iff v(\lambda') \leq v(\mu'). \end{aligned}$$

□

We shall now extend the dominance partial ordering (2.6.4) on L'_{++} to a partial ordering on L', as follows: for $\lambda', \mu' \in L'$,

(2.7.5) $\lambda' \geq \mu'$ *if and only if either* (i) $\lambda'_+ > \mu'_+$, *or* (ii) $\lambda'_+ = \mu'_+$ *and* $v(\lambda') \leq v(\mu')$ (*or equivalently* (2.7.4) $\bar{v}(\lambda') \geq \bar{v}(\mu')$).

Observe that for this ordering, in a given W_0-orbit the *antidominant* weight is highest.

(2.7.6) *Let $v, w \in W_0$. If $v \leq w$ then $v\lambda' - w\lambda' \in Q^\vee_+$ for all $\lambda \in L'_{++}$.*

Proof We may assume that $w = vs_\alpha$ where $\alpha \in R^+$ and $v\alpha \in R^+$. Hence

$$v\lambda' - w\lambda' = v(\lambda' - s_\alpha\lambda') = \langle\lambda', \alpha\rangle v\alpha^\vee$$

which is in $Q^\vee_+$ because $\langle\lambda', \alpha\rangle \geq 0$. □

(2.7.7) *Let $\lambda', \mu' \in L'$ lie in the same W_0-orbit. If $\lambda' \geq \mu'$ then $\mu' - \lambda' \in Q^\vee_+$.*

Proof We have $\mu' - \lambda' = \bar{v}(\mu')\lambda'_+ - \bar{v}(\lambda')\lambda'_+$. Hence the result follows from (2.7.6). □

Remark The converse of (2.7.7) is in general false, if the rank of R is greater than 2. (For example, if R is of type A_3 let $\lambda' = -\varepsilon_2 - 2\varepsilon_3 + 3\varepsilon_4$, $\mu' = 3\varepsilon_1 - 2\varepsilon_2 - \varepsilon_3$, in the notation of (1.3.1). Here $\lambda'_+ = \mu'_+ = 3\varepsilon_1 - \varepsilon_3 - 2\varepsilon_4$ and $\mu' - \lambda' = 3\varepsilon_1 - \varepsilon_2 + \varepsilon_3 - 3\varepsilon_4 = 3\alpha_1 + 2\alpha_2 + 3\alpha_3 \in Q^\vee_+$. But $v(\lambda') = s_1 s_2 s_1$ and $v(\mu') = s_3 s_2 s_1$ are incomparable for the Bruhat order.)

(2.7.8) *Let λ', μ' lie in the same W_0-orbit. Then the following are equivalent*: (a) $\lambda' \geq \mu'$; (b) $-\mu' \geq -\lambda'$; (c) $w_0\mu' \geq w_0\lambda'$.

This follows from (2.7.1). □

(2.7.9) *Let $\lambda' \in L', \alpha \in R^+$. Then $<\lambda', \alpha> > 0$ if and only if $s_\alpha\lambda' > \lambda'$.*

Proof Let $\mu' = s_\alpha\lambda'$. if $<\lambda', \alpha> > 0$ then $\alpha \in s(v(\lambda'))$ by (2.7.2), hence $v(\lambda')s_\alpha < v(\lambda')$ by (2.3.3). Now $v(\lambda')s_\alpha$ takes μ' to $\lambda'_- = \mu'_-$, hence $v(\lambda')s_\alpha \geq v(\mu')$ It follows that $u(\lambda') > u(\mu')$, i.e., $\mu' > \lambda'$.

If on the other hand $<\lambda', \alpha> < 0$, we have $<\mu', \alpha> > 0$ and hence $\lambda' > \mu'$ by the previous paragraph. Finally, if $<\lambda', \alpha> = 0$ then $\mu' = \lambda'$. □

(2.7.10) (i) *Let $\lambda' \in L'$*, let $v(\lambda') = s_{i_1} \cdots s_{i_p}$ *be a reduced expression, and let* $\lambda'_r = s_{i_{r+1}\cdots s_{i_p}}(\lambda')$, *for* $0 \leq r \leq p$. *Then*

$$\lambda'_- = \lambda'_0 > \lambda'_1 > \cdots > \lambda'_p = \lambda'.$$

(ii) *Let* $\bar{v}(\lambda') = s_{j_q} \cdots s_{j_1}$ *be a reduced expression, and let* $\mu'_r = s_{j_{r+1}} \cdots s_{j_q}(\lambda')$, *for* $0 \leq r \leq q$. *Then*

$$\lambda'_+ = \mu'_0 < \mu'_1 < \cdots < \mu'_q = \lambda'.$$

Proof (i) Let $\beta_r = s_{i_p} \cdots s_{i_{r+1}}(\alpha_{i_r})$ for $1 \leq r \leq p$, so that $s(v(\lambda')) = \{\beta_1, \ldots, \beta_p\}$ by (2.2.9). Hence $<\lambda'_r, \alpha_{i_r}> = <\lambda', \beta_r> > 0$ by (2.7.2), and therefore $\lambda'_{r-1} = s_{i_r}\lambda'_r > \lambda'_r$ by (2.7.9).
(ii) Let $\gamma_r = s_{j_q} \cdots s_{j_{r+1}}(\alpha_{j_r})$ for $1 \leq r \leq q$, so that $s(\bar{v}(\lambda')^{-1}) = \{\gamma_1, \ldots, \gamma_q\}$ by (2.2.9). Hence $<\mu'_r, \alpha_{j_r}> = <\lambda', \gamma_r> < 0$ by (2.7.2), and therefore $\mu'_{r-1} = s_{j_r}\mu'_r < \mu'_r$ by (2.7.9). □

(2.7.11) *Let $v, w \in W$ and let $v(0) = \lambda'$, $w(0) = \mu'$. Then*

$$v \leq w \Rightarrow u(\lambda') \leq u(\mu') \Rightarrow \lambda' \leq \mu'.$$

Proof We have $w = u(\mu')w'$, where $w' \in W_0$ and $l(w) = l(u(u')) + l(w')$, by (2.4.9). Since $v \leq w$, it follows that $v = v_1v_2$, where $v_1 \leq u(\mu')$ and $v_2 \leq w'$, so that $v_2 \in W_0$. Hence $v_1(0) = v(0) = \lambda'$, and so $v_1 \geq u(\lambda')$ Consequently $u(\lambda') \leq u(\mu')$.

We shall next show that $v \leq w$ implies $\lambda' \leq \mu'$. For this purpose we may assume that $v = ws_a$ where $a \in S(w)$, by (2.3.3). Let $a = \alpha + rc$ and let

$w = t(\mu')w'$ where $w' \in W_0$. From (2.1.4) we have

$$wa = \beta + (r - <\mu', \beta>)c$$

where $\beta = w'\alpha$, and

$$\lambda' = v(0) = t(\mu')ws_a(0) = \mu' - r\beta^\vee.$$

Since $a \in S(w)$ we must have

(1) $$\chi(\alpha) \le r < \chi(\beta) + <\mu', \beta>$$

from which it follows that $<\mu', \beta> \ge 0$. If $<\mu', \beta> = 0$ then $r = 0$ and $\lambda' = \mu'$. If $<\mu', \beta> > 0$ and $0 < r < <\mu', \beta>$, then λ' lies in the interior of the line segment $[\mu', s_\beta\mu']$, so that $\Sigma(\lambda')$ is strictly contained in $\Sigma(\mu')$ and therefore $\lambda' < \mu'$ by (2.6.3). Finally, if $r = <\mu', \beta>$, so that $\lambda' = s_\beta\mu'$, we must have $\chi(\beta) = 1$ by (1) above, hence $\beta \in R^-$ and therefore $s_\beta\mu' < \mu'$ by (2.7.9). Hence $\lambda' < \mu'$ in this case also.

Finally, by taking $v = u(\lambda')$ and $w = u(\mu')$, it follows that $u(\lambda') \le u(\mu')$ implies $\lambda' \le \mu'$. □

(2.7.12) *Let* $w \in W$, $\mu' \in L'$. *If* $w < u(\mu')$ *then* $w(0) < \mu'$.

Proof Let $w(0) = \lambda'$. Then $\lambda' \le \mu'$ by (2.7.11), and $\lambda' \ne \mu'$, hence $\lambda' < \mu'$. □

(2.7.13) *Let* $\lambda' \in L', i \in I$. *Then* $s_i\lambda' > \lambda'$ *if and only if* $a_i(\lambda') > 0$.

Proof This follows from (2.7.9) if $i \ne 0$. If $i = 0$ and $a_0(\lambda') = r > 0$ let $\mu' = s_0\lambda' = \lambda' - a_0(\lambda')\alpha_0^\vee = \lambda' + r\varphi^\vee$, and $s_\varphi\mu' = \lambda' - \varphi^\vee$, so that λ' lies in the interior of the segment $[\mu', s_\varphi\mu']$ and therefore $\lambda' < \mu'$. Finally, if $a_0(\lambda') < 0$, interchange λ' and μ'. □

2.8 The functions $r_{k'}, r'_k$

Let S, S' be as in §1.4 and let k be a W-labelling of S as defined in §1.5, and k' the dual labelling of S'. For each $\lambda' \in L'$ let $u(\lambda')$ be the shortest element of the coset $t(\lambda')W_0$, as in §2.4, and define

(2.8.1) $$r'_k(\lambda') = u(\lambda')(-\rho'_k)$$

where ρ'_k is given by (1.5.2).

Dually, if $\lambda \in L$ let $u'(\lambda) \in W'$ be the shortest element of the coset $t(\lambda)W_0$, and define

(2.8.1′) $$r_{k'}(\lambda) = u'(\lambda)(-\rho_{k'}).$$

Then we have

(2.8.2) $$r'_k(\lambda') = \lambda' + \frac{1}{2}\sum_{\alpha \in R^+} \eta(<\lambda', \alpha>)k(\alpha'^\vee)\alpha',$$

(2.8.2′) $$r_{k'}(\lambda) = \lambda + \frac{1}{2}\sum_{\alpha \in R^+} \eta(<\lambda, \alpha'>)k'(\alpha^\vee)\alpha,$$

where for $x \in \mathbb{R}$

(2.8.3) $$\eta(x) = \begin{cases} 1 & \text{if } x > 0, \\ -1 & \text{if } x \leq 0. \end{cases}$$

Proof Since $u(\lambda') = t(\lambda')v(\lambda')^{-1}$, we have

$$\begin{aligned} r'_k(\lambda') &= \lambda' - v(\lambda')^{-1}(\rho'_k) \\ &= \lambda' - \frac{1}{2}\sum_{\alpha \in R^+} \sigma(v(\lambda')\alpha)k(\alpha'^\vee)\alpha' \end{aligned}$$

by (1.5.3), and $\sigma(v(\lambda')\alpha) = -\eta(<\lambda', \alpha>)$ by (2.4.4). □

(2.8.4) *Let* $\lambda' \in L'$.
(i) *If* $u_j \in \Omega$ *then* $r'_k(u_j\lambda') = u_j(r'_k(\lambda'))$.
(ii) *If* $i \in I$ *and* $\lambda' \neq s_i\lambda'$, *then* $s_i(r'_k(\lambda')) = r'_k(s_i\lambda')$.
(iii) *If* $i \in I$ *and* $\lambda' = s_i\lambda'$, *then* $s_i(r'_k(\lambda')) = r'_k(\lambda') + k(\alpha_i'^\vee)\alpha'_i$.

Proof (i) follows from (2.4.12), and (ii) from (2.4.14) (i).
(iii) From (2.4.14) (iii) we have $s_iu(\lambda') = u(\lambda')s_j$ and $v(\lambda')\alpha_i = \alpha_j$ for some $j \neq 0$, so that

$$\begin{aligned} s_i(r'_k(\lambda')) &= s_iu(\lambda')(-\rho'_k) = u(\lambda')s_j(-\rho'_k) \\ &= u(\lambda')(-\rho'_k + k(\alpha_j'^\vee)\alpha'_j) \\ &= u(\lambda')(-\rho'_k) + k(\alpha_j'^\vee)v(\lambda')^{-1}\alpha'_j \\ &= r'_k(\lambda') + k(\alpha_i'^\vee)\alpha'_i. \end{aligned}$$ □

For the rest of this section we shall assume that $k(\alpha'^\vee) \geq 0$ for each $\alpha \in R$.

(2.8.5) *The mapping* $r'_k; L' \to V$ *is injective.*

Proof Let λ', μ' be such that $r'_k(\lambda') = r'_k(\mu')$. We have

$$v(\lambda')(r'_k(\lambda')) = v(\lambda')u(\lambda')(-\rho'_k) = \lambda'_- - \rho'_k$$

by (2.4.3), where λ'_- is the antidominant element of the orbit $W_0\lambda'$. Hence (as the labels are all ≥ 0) $\lambda'_- - \rho'_k$ is the antidominant element of the orbit $W_0 r'_k(\lambda')$. So if $r'_k(\lambda') = r'_k(\mu')$ we must have $\lambda'_- - \rho'_k = \mu'_- - \rho'_k$, hence $\lambda'_- = \mu'_-$ and $v(\lambda') = v(\mu')$, whence $\lambda' = \mu'$. □

(2.8.6) *Let $\lambda' \in L'$. If $s_i\lambda' = \lambda'$ for some $i \in I$, then $s_i(r'_k(\lambda')) \notin r'_k(L')$.*

Proof Suppose that $s_i(r'_k(\lambda')) = r'_k(\mu')$ for some $\mu' \in L'$. Then as in (2.8.5) we have

$$s_i v(\lambda')^{-1}(\lambda'_- - \rho'_k) = v(\mu')^{-1}(\mu'_- - \rho'_k)$$

from which we conclude that $\lambda'_- = \mu'_-$ and $s_i v(\lambda')^{-1} = v(\mu')^{-1}$. Consequently $\mu' = v(\mu')^{-1}\mu'_- = s_i v(\lambda')^{-1}\lambda'_- = s_i\lambda' = \lambda'$. But $s_i(r'_k(\lambda')) \neq r'_k(\lambda')$ by (2.8.4) (iii). □

Notes and references

The extended affine Weyl groups occur in [B1], p. 176, and probably earlier. The elements $u(\lambda')$, $v(\lambda')$ were defined by Cherednik in [C2], and the partial order on L' was introduced by Heckman (see [O4], Def. 2.4).

3

The braid group

3.1 Definition of the braid group

We retain the notation of Chapter 2. The *braid group* $\mathfrak{B}$ of the extended affine Weyl group W is the group with generators $T(w)$, $w \in W$, and relations

(3.1.1) $$T(v)T(w) = T(vw) \quad \text{if} \quad l(v) + l(w) = l(vw).$$

There is an obvious surjective homomorphism

(3.1.2) $$f : \mathfrak{B} \to W$$

such that $f(T(w)) = w$ for each $w \in W$.

We shall write

$$T_i = T(s_i), \quad U_j = T(u_j)$$

for $i \in I$ and $j \in J$.

Let i, j be distinct elements of I such that $s_i s_j$ has finite order m_{ij} in W. Then we have

$$s_i s_j s_i \cdots = s_j s_i s_j \cdots$$

with m_{ij} factors on either side. Since both sides are reduced expressions, it follows from (3.1.1) that

(3.1.3) $$T_i T_j T_i \cdots = T_j T_i T_j \cdots$$

with m_{ij} factors on either side.

These relations (3.1.3) are called the *braid relations*.

Next, let $j, k \in J$. Then $u_j u_k = u_{j+k}$ (2.5.6), and all three terms have length zero, so that

(3.1.4) $$U_j U_k = U_{j+k}.$$

Finally, let $i \in I$ and $j \in J$. Then $u_j(a_i) = a_{i+j}$ (2.5.7), so that $u_j s_i = s_{i+j} u_j$, and therefore $U_j T_i = T_{i+j} U_j$ by (3.1.1), i.e.,

$$U_j T_i U_j^{-1} = T_{i+j}. \tag{3.1.5}$$

(3.1.6) *The braid group* $\mathfrak{B}$ *is generated by the* T_i $(i \in I)$ *and the* U_j $(j \in J)$ *subject to the relations* (3.1.3), (3.1.4), (3.1.5).

Proof Each $w \in W$ may be written in the form $w = u_j s_{i_1} \cdots s_{i_p}$, where $i_1, \dots, i_p \in I$, $j \in J$ and $p = l(w)$. It follows from (3.1.1) that $T(w) = U_j T_{i_1} \cdots T_{i_p}$, and hence that the T_i and the U_j generate $\mathfrak{B}$.

Now let $\mathfrak{B}'$ be a group with generators T_i $(i \in I)$ and U_j $(j \in J)$ and relations (3.1.3), (3.1.4), (3.1.5). For w as above, define

$$T'(w) = U_j T_{i_1} \cdots T_{i_p}.$$

The braid relations (3.1.3) guarantee that this definition is unambiguous. Next, if $w' = u_k s_{j_1} \cdots s_{j_q}$ is a reduced expression (so that $q = l(w')$) we have

$$T'(w') = U_k T_{j_1} \cdots T_{j_q}$$

and

$$\begin{aligned} ww' &= u_j s_{i_1} \cdots s_{i_p} u_k s_{j_1} \cdots s_{j_q} \\ &= u_{j+k} s_{i_1 - k} \cdots s_{i_p - k} s_{j_1} \cdots s_{j_q}. \end{aligned}$$

If now $l(w) + l(w') = l(ww')$, we have

$$T'(ww') = U_{j+k} T_{i_1 - k} \cdots T_{i_p - k} T_{j_1} \cdots T_{j_q}$$

which is equal to $T'(w)T'(w')$ by use of the relations (3.1.4) and (3.1.5). It follows that the relations (3.1.1) are consequences of (3.1.3)–(3.1.5), and hence $\mathfrak{B}'$ is isomorphic to $\mathfrak{B}$. □

(3.1.7) *Let* $w \in W, i \in I$. *Then*

$$T(ws_i) = T(w) T_i^{\sigma(wa_i)},$$

$$T(s_i w) = T_i^{\sigma(w^{-1}a_i)} T(w)$$

where (2.2.7) $\sigma(a) = +1$ *or* -1 *according as* $a \in S^+$ *or* $a \in S^-$.

Proof This follows from (2.2.8). □

(3.1.8) *Let* $w \in W$ *and* $i, j \in I$. *If* $ws_i = s_j w$ *then* $T(w)T_i = T_j T(w)$.

Proof We have $s_{wa_i} = ws_i w^{-1} = s_j$, hence $wa_i = \varepsilon a_j$ where $\varepsilon = \pm 1$. Hence, from (3.1.7),

$$T(w)T_i^{\varepsilon} = T(ws_i) = T(s_j w) = T_j^{\varepsilon} T(w)$$

and therefore $T(w)T_i = T_j T(w)$. □

(3.1.9) *Let* $u, v \in W$ *and let* $u^{-1}v = u_j s_{i_1} \cdots s_{i_p}$ *be a reduced expression (so that* $l(u^{-1}v) = p$*). Let* $b_r = u_j s_{i_1} \cdots s_{i_{r-1}}(a_{i_r})$ *for* $1 \le r \le p$. *Then*

$$T(u)^{-1}T(v) = U_j T_{i_1}^{\varepsilon_1} \cdots T_{i_p}^{\varepsilon_p}$$

where $\varepsilon_r = \sigma(ub_r)\,(1 \le r \le p)$.

Proof This is by induction on p. If $p = 0$ we have $v = uu_j$ and therefore $T(v) = T(u)U_j$ by (2.2.5). If $p \ge 1$ we have

$$u^{-1}vs_{i_p} = u_j s_{i_1} \cdots s_{i_{p-1}}$$

and hence by the inductive hypothesis

$$T(u)^{-1}T(vs_{i_p}) = U_j T_{i_1}^{\varepsilon_1} \cdots T_{i_{p-1}}^{\varepsilon_{p-1}}$$

with $\varepsilon_1, \ldots, \varepsilon_{p-1}$ as above. Since

$$va_{i_p} = uu_j s_{i_1} \cdots s_{i_p}(a_{i_p}) = -ub_p,$$

it follows from (3.1.7) that $T(vs_{i_p}) = T(v)T_{i_p}^{-\varepsilon_p}$. Hence

$$T(u)^{-1}T(v) = U_j T_{i_1}^{\varepsilon_1} \cdots T_{i_p}^{\varepsilon_p}.$$ □

3.2 The elements $Y^{\lambda'}$

Let $\lambda' \in L'$. If λ' is dominant we define

(3.2.1) $$Y^{\lambda'} = T(t(\lambda')).$$

If λ' and μ' are both dominant, we have $l(t(\lambda' + \mu')) = l(t(\lambda')) + l(t(\mu'))$ by (2.4.2), and hence

(3.2.2) $$Y^{\lambda'+\mu'} = Y^{\lambda'}Y^{\mu'} = Y^{\mu'}Y^{\lambda'}.$$

Now let λ' be any element of L'. We can write $\lambda' = \mu' - \nu'$, where $\mu', \nu' \in L'$ are dominant, and we define

$$Y^{\lambda'} = Y^{\mu'}(Y^{\nu'})^{-1}. \tag{3.2.3}$$

This definition is unambiguous, because if also $\lambda' = \mu'_1 - \nu'_1$ with μ'_1, ν'_1 dominant, we have $\mu' + \nu'_1 = \mu'_1 + \nu'$ and therefore $Y^{\nu'_1}Y^{\mu'} = Y^{\mu'_1}Y^{\nu'}$ by (3.2.2). The relation (3.2.2) now holds for all $\lambda', \mu' \in L'$, and the set

$$Y^{L'} = \{Y^{\lambda'} : \lambda' \in L'\}$$

is a commutative subgroup of $\mathfrak{B}$, isormorphic to L'. For the homomorphism $f : \mathfrak{B} \to W$ (3.1.2) maps $Y^{L'}$ onto $t(L')$.

(3.2.4) *Let $\lambda' \in L'$ and $i \in I_0$ be such that $<\lambda', \alpha_i> = 0$ or 1. Then*

$$T_i^{\varepsilon} Y^{s_i\lambda'} T_i = Y^{\lambda'}$$

where

$$\varepsilon = \begin{cases} +1 & \text{if } <\lambda', \alpha_i> = 1, \\ -1 & \text{if } <\lambda', \alpha_i> = 0. \end{cases}$$

When $<\lambda', \alpha_i> = 0$, so that $s_i\lambda' = \lambda'$, (3.2.4) says that

$$T_i Y^{\lambda'} = Y^{\lambda'} T_i. \tag{3.2.5}$$

When $<\lambda', \alpha_i> = 1$ we have $s_i\lambda' = \lambda' - \alpha_i^\vee$, and (3.2.4) takes the form

$$T_i Y^{\lambda' - \alpha_i^\vee} T_i = Y^{\lambda'}. \tag{3.2.6}$$

Proof We begin with (3.2.5). We may write $\lambda' = \mu' - \nu'$ with μ', ν' both dominant and $<\mu', \alpha_i> = <\nu', \alpha_i> = 0$. Then s_i commutes with $t(\mu')$ and $t(\nu')$, hence by (3.1.8) T_i commutes with both $Y^{\mu'}$ and $Y^{\nu'}$, hence also with $Y^{\lambda'}$.

Next, to prove (3.2.6), suppose first that λ' is dominant. Then $\mu' = \lambda' + s_i\lambda'$ is also dominant, and $<\mu', \alpha_i> = 0$. Let $w = t(\lambda')s_i t(\lambda') = s_i t(\mu')$. If $l(t(\lambda')) = p$ we have

$$l(t(\mu')) = 2p - 2, \quad l(w) = 2p - 1, \quad l(t(\lambda')s_i) = p - 1,$$

by use of the length formula (2.4.1). Hence

$$T_i Y^{s_i\lambda' + \lambda'} = T_i Y^{\mu'} = T(w) = T(t(\lambda')s_i)T(t(\lambda')) = Y^{\lambda'} T_i^{-1} Y^{\lambda'}$$

which gives (3.2.6) for λ' dominant. If now λ' is not dominant, let $\nu' = \lambda' - \pi_i'$, so that $<\nu', \alpha_i> = 0$. Then we have

$$Y^{\lambda'} = Y^{\pi_i'} Y^{\nu'} = T_i Y^{s_i \pi_i'} T_i Y^{\nu'} = T_i Y^{s_i \pi_i' + \nu'} T_i = T_i Y^{s_i \lambda'} T_i,$$

since T_i commutes with $Y^{\nu'}$ by (3.2.5). □

(3.2.7) *Remark* If (3.2.6) is not vacuous, that is to say if there exists $\lambda' \in L'$ such that $<\lambda', \alpha_i> = 1$, then (3.2.5) is a consequence of (3.2.6). (For if $<\mu', \alpha_i> = 0$ then $<\lambda' + \mu', \alpha_i> = 1$, and hence

$$T_i Y^{\lambda' + \mu' - \alpha_i^\vee} T_i = Y^{\lambda' + \mu'} = T_i Y^{\lambda' - \alpha_i^\vee} T_i Y^{\mu'}$$

giving $Y^{\mu'} T_i = T_i Y^{\mu'}$.)

However, there is one case in which (3.2.6) is vacuous, namely (2.1.6) when R is of type C_n, $L' = Q^\vee$ and α_i is the long simple root of R. In that case $<\lambda', \alpha_i>$ is an even integer for all $\lambda' \in L'$.

As in Chapter 2, let φ be the highest root of R and recall (2.5.2) that $u_j = t(\pi_j')v_j^{-1}$ for $j \in J$. We have then

(3.2.8) $$T_0 = Y^{\varphi^\vee} T(s_\varphi)^{-1},$$

(3.2.9) $$U_j = Y_j' T(v_j)^{-1}$$

for $j \in J$, where $Y_j' = Y^{\pi_j'}$. (In particular, $U_0 = 1$.)

Proof We have $s_0 s_\varphi = t(\varphi^\vee)$, and $s_\varphi(a_0) = \varphi + c \in S^+$, so that by (3.1.7) $Y^{\varphi^\vee} = T_0 T(s_\varphi)$, which gives (3.2.8).

Next, $t(\pi_j') = u_j v_j$ and $l(u_j) = 0$, so that $Y_j' = U_j T(v_j)$, giving (3.2.9). □

(3.2.10) *Let $\lambda' \in L'$ and let $u(\lambda') = u_j s_{i_1} \cdots s_{i_q}$ be a reduced expression. Then*

$$Y^{\lambda'} = U_j T_{i_1}^{\varepsilon_1} \cdots T_{i_q}^{\varepsilon_q} T(v(\lambda'))$$

where each exponent ε_r is ± 1.

Proof Let $\lambda' = \mu' - \nu'$ with $\mu', \nu' \in L'$ both dominant, and let $v(\lambda') = s_{i_{q+1}} \cdots s_{i_p}$ be a reduced expression. Take $u = t(\nu')$ and $v = t(\mu')$ in (3.1.9): we have

$$u^{-1} v = t(\lambda') = u(\lambda') v(\lambda') = u_j s_{i_1} \cdots s_{i_p}$$

which is a reduced expression, since $l(u(\lambda')) + l(v(\lambda')) = l(t(\lambda'))$ by (2.4.5). Hence by (3.1.9)

$$Y^{\lambda'} = T(u)^{-1}T(v) = U_j T_{i_1}^{\varepsilon_1} \cdots T_{i_p}^{\varepsilon_p}$$

where $\varepsilon_r = \sigma(t(\nu')b_r)$ and $b_r = u_j s_{i_1} \cdots s_{i_{r-1}}(a_{i_r})$. We have to show that $\varepsilon_r = +1$ for each $r > q$, i.e. that $t(\nu')b_r \in S^+$.

If $r > q$ then $i_r \neq 0$, hence $a_{i_r} = \alpha_{i_r}$ and therefore

$$\begin{aligned} b_r &= u(\lambda')s_{i_{q+1}} \cdots s_{i_{r-1}}(\alpha_{i_r}) \\ &= t(\lambda')s_{i_p} \cdots s_{i_r}(\alpha_{i_r}) = -t(\lambda')\beta_r, \end{aligned}$$

where $\beta_r = s_{i_p} \cdots s_{i_{r+1}}(\alpha_{i_r}) \in S(v(\lambda'))$ by (2.2.9), so that $<\lambda', \beta_r> > 0$ by (2.4.4). Hence

$$\begin{aligned} t(\mu')b_r &= -t(\lambda' + \nu')\beta_r = -t(\mu')\beta_r \\ &= -\beta_r + <\mu', \beta_r>c \end{aligned}$$

and $<\mu', \beta_r> = <\lambda', \beta_r> + <\nu', \beta_r> \geq 1$, since $\beta_r \in R^+$ and $\nu' \in L'$ is dominant. Hence $t(\nu')b_r \in S^+$, as required. □

3.3 Another presentation of $\mathfrak{B}$

Let $\mathfrak{B}_0$ be the subgroup of $\mathfrak{B}$ generated by the T_i, $i \neq 0$. In this section we shall show that

(3.3.1) *$\mathfrak{B}$ is generated by $\mathfrak{B}_0$ and $Y^{L'}$, subject to the relations* (3.2.4).

It follows from (3.1.6), (3.2.8) and (3.2.9) that $\mathfrak{B}$ is generated by $\mathfrak{B}_0$ and $Y^{L'}$. Let $\mathfrak{B}'$ denote the group generated by $\mathfrak{B}_0$ and $Y^{L'}$ subject to the relations (3.2.4). In $\mathfrak{B}'$ we *define* elements T_0, U_j $(j \in J)$ by means of (3.2.8) and (3.2.9). We have then to show that with these definitions the relations (3.1.3)–(3.1.5) hold in $\mathfrak{B}'$. We remark that (3.1.7)–(3.1.9), restricted to elements of W_0, hold in $\mathfrak{B}'$ (because they hold in $\mathfrak{B}_0$).

(3.3.2) *Let $\lambda' \in L'$ and $w \in W_0$ be such that $<\lambda', \alpha> = 0$ or 1 for all $\alpha \in S(w^{-1})$. Then in $\mathfrak{B}'$ we have*

$$T(v(\lambda')w)^{-1}T(v(\lambda'))Y^{-\lambda'}T(w) = Y^{-w^{-1}\lambda'}.$$

Proof We shall apply (3.1.9) with $u = v(\lambda')w$ and $v = v(\lambda')$, so that $w = v^{-1}u = s_{i_p} \cdots s_{i_1}$ and therefore $T(w) = T_{i_p} \cdots T_{i_1}$. In this way we obtain

(1) $$T(v(\lambda')w)^{-1}T(v(\lambda'))Y^{-\lambda'}T(w) = T_{i_1}^{\varepsilon_1} \cdots T_{i_p}^{\varepsilon_p} Y^{-\lambda'} T_{i_p} \cdots T_{i_1}$$

where, in the notation of (3.1.9), $\varepsilon_r = \sigma(ub_r)$ and

$$\begin{aligned} ub_r &= v(\lambda')ws_{i_1} \cdots s_{i_{r-1}}(\alpha_{i_r}) = v(\lambda')s_{i_p} \cdots s_{i_r}(\alpha_{i_r}) \\ &= -v(\lambda')\gamma_r, \end{aligned}$$

where $\gamma_r = s_{i_p} \cdots s_{i_{r+1}}(\alpha_{i_r})$, so that $\{\gamma_1, \ldots, \gamma_p\} = S(w^{-1})$ by (2.2.9). If $<\lambda', \gamma_r> = 1$ then $\gamma_r \in S(v(\lambda'))$ by (2.4.4) and hence $\varepsilon_r = 1$. If on the other hand $<\lambda', \gamma_r> = 0$ then $\gamma_r \notin S(v(\lambda'))$ and so $\varepsilon_r = -1$.

Now let $\lambda'_r = s_{i_{r+1}} \cdots s_{i_p}(\lambda')$ for $0 \le r \le p$, so that $\lambda'_p = \lambda'$ and $\lambda'_0 = w^{-1}\lambda'$. To complete the proof it is enough to show that $T_{i_r}^{\varepsilon_r} Y^{-\lambda'_r} T_{i_r} = Y^{-\lambda'_{r-1}}$ for $1 \le r \le p$; now this follows from (3.2.4), since $\lambda'_{r-1} = s_{i_r}\lambda'_r$ and $<\lambda'_r, \alpha_{i_r}> = <\lambda', \gamma_r> = 1$ or 0 according as $\varepsilon_r = +1$ or -1. □

We shall apply (3.3.2) when $\lambda' = \pi'_j$ $(j \in J)$ and when $\lambda' = \varphi^\vee$. We have $<\pi'_j, \alpha> = 0$ or 1 for all $\alpha \in R^+$, and $<\varphi^\vee, \alpha> = 0$ or 1 for all $\alpha \in R^+$ except $\alpha = \varphi$. When $\lambda' = \pi'_j$, we have $v(\lambda') = v_j$ (2.5.2) and when $\lambda' = \varphi^\vee$, $v(\lambda') = s_\varphi$ by (2.4.11). Hence (3.3.2) gives

(3.3.3) $$U_j = T(w)Y^{w^{-1}\pi'_j}T(v_j w)^{-1}$$

for all $w \in W_0$, and

(3.3.4) $$T_0 = T(w)Y^{w^{-1}\varphi^\vee}T(s_\varphi w)^{-1}$$

for all $w \in W_0$ such that $w^{-1}\varphi \in R^+$. In particular,

(3.3.5) $$T_0 = T(w)Y^{\alpha_i^\vee}T_i^{-1}T(w)^{-1}$$

if $w \in W_0$ is such that $\varphi = w\alpha_i$, $i \ne 0$.

In (3.3.4) let $v = s_\varphi w$, so that $v^{-1}\varphi = -w^{-1}\varphi \in R^-$. We have then

$$T_0 = T(s_\varphi v)Y^{-v^{-1}\varphi^\vee}T(v)^{-1}$$

and therefore, for all $w \in W_0$,

(3.3.6) $$T_0^\varepsilon = T(w)Y^{w^{-1}\varphi^\vee}T(s_\varphi w)^{-1}$$

where $\varepsilon = \sigma(w^{-1}\varphi)$.

In particular, let $w = v_j$ $(j \in J)$. Then $w^{-1}\varphi = -v_j^{-1}\alpha_0 = -\alpha_j$ by (2.5.8), so that $T(s_\varphi w) = T(ws_j) = T(w)T_j^{-1}$ by (3.1.7), and therefore

(3.3.7) $$T_0 = T(v_j)T_j^{-1}Y^{\alpha_j^\vee}T(v_j)^{-1}.$$

We shall now show that the relations (3.1.3)–(3.1.5) hold in $\mathfrak{B}'$.

Proof of (3.1.3). We need only consider the braid relations that involve T_0, which are

(a) $T_0 T_i = T_i T_0$ if $<\varphi^\vee, \alpha_i> = 0$,
(b) $T_0 T_i T_0 = T_i T_0 T_i$ if $<\varphi^\vee, \alpha_i> = <\varphi, \alpha_i^\vee> = 1$,
(c) $T_0 T_i T_0 T_i = T_i T_0 T_i T_0$ if $<\varphi^\vee, \alpha_i> = 1$, $<\varphi, \alpha_i^\vee> = 2$.

(The list in §1.3 shows that $<\varphi, \alpha_i^\vee> = 3$ does not occur.)

(a) If $<\varphi^\vee, \alpha_i> = 0$ then T_i commutes with $Y^{\varphi^\vee}$ by (3.2.5) and with $T(s_\varphi)$ by (3.1.8), hence with T_0.

(b) If $<\varphi^\vee, \alpha_i> = <\varphi, \alpha_i^\vee> = 1$, then $s_\varphi(\alpha_i) = \alpha_i - \varphi \in R^-$, hence by (3.1.7)

(1) $$T(s_i s_\varphi) = T_i^{-1} T(s_\varphi) = T_i^{-1} T_0^{-1} Y^{\varphi^\vee}.$$

Let $w = s_i s_\varphi s_i = w^{-1}$. Since $s_i s_\varphi(\alpha_i) = -\varphi \in R^-$, we have $T(w) = T(s_i s_\varphi) T_i^{-1} = T_i^{-1} T_0^{-1} Y^{\varphi^\vee} T_i^{-1}$, and hence by (3.2.6)

(2) $$T(w) = T_i^{-1} T_0^{-1} T_i Y^{s_i \varphi^\vee}.$$

Next, since $w\alpha_i = \varphi$, (3.3.4) gives

(3) $$T(w) = T_0 T(s_\varphi w) Y^{-w\varphi^\vee}.$$

Since $s_\varphi w = s_i s_\varphi$, it follows from (1) and (3) that

(4) $$T(w) = T_0 T_i^{-1} T_0^{-1} Y^{\varphi^\vee - \alpha_i^\vee}.$$

Comparison of (2) and (4) now shows that $T_i^{-1} T_0^{-1} T_i - T_0 T_i^{-1} T_0^{-1}$, hence that $T_0 T_i T_0 = T_i T_0 T_i$.

(c) Now suppose that $<\varphi^\vee, \alpha_i> = 1$ and $<\varphi, \alpha_i^\vee> = 2$. The relations (1)–(3) above still hold, since now $s_i s_\varphi(\alpha_i) = \alpha_i - \varphi \in R^-$, and $w^{-1}\varphi = \varphi \in R^+$. Let $v = s_\varphi w = (s_\varphi s_i)^2 = (s_i s_\varphi)^2$. From (2) and (3) we have

(5) $$T(v) = T_0^{-1} T_i^{-1} T_0^{-1} T_i Y^{\varphi^\vee + s_i \varphi^\vee}.$$

We need one more relation, which we obtain by taking $w = s_i s_\varphi$ in (3.3.4): this is legitimate, since $(s_i s_\varphi)^{-1}\varphi = s_i \varphi \in R^+$. We obtain

(6) $$T(s_\varphi s_i s_\varphi) = T_0^{-1} T(s_i s_\varphi) Y^{s_i \varphi^\vee}.$$

Now $v\alpha_i = -\alpha_i$ and therefore $T(s_\varphi s_i s_\varphi) = T(s_i v) = T_i^{-1} T(v)$, so that (1) and (6) give

(7) $$T(v) = T_i T_0^{-1} T_i^{-1} T_0^{-1} Y^{\varphi^\vee + s_i \varphi^\vee}.$$

Comparison of (5) and (7) now shows that $T_0 T_i T_0 T_i = T_i T_0 T_i T_0$. □

Proof of (3.1.4). Let $j, k \in J$. We may assume that $j \neq 0$ and $k \neq 0$, since $U_0 = 1$. Take $w = v_k$ in (3.3.3); since $v_j v_k = v_{j+k}$ and $v_k^{-1}\pi'_j = \pi'_{j+k} - \pi'_k$ by (2.5.9) and (2.5.10), we obtain

$$U_j = T(v_k)Y_k'^{-1}Y'_{j+k}T(v_{j+k})^{-1} = U_k^{-1}U_{j+k}$$

and hence $U_k U_j = U_{j+k}$. □

Proof of (3.1.5). We have to show that $U_j T_i U_j^{-1} = T_{i+j}$ for $i \in I$ and $j \in J$. As before, we may assume that $j \neq 0$.
(a) Suppose that $i \neq 0$ and $i + j \neq 0$. Then $v_j^{-1}\alpha_i = \alpha_{i+j}$ (2.5.8) and $<\pi'_j, \alpha_{i+j}> = 0$, hence by (3.1.8) and (3.2.5)

$$U_j T_i U_j^{-1} = Y'_j T(v_j)^{-1} T_i T(v_j) Y_j^{-1} = Y'_j T_{i+j} Y_j'^{-1} = T_{i+j}.$$

(b) Suppose now that $i = 0$. By (3.3.7) we have

$$\begin{aligned} T_0 &= T(v_j)T_j^{-1}Y^{\alpha_j^\vee}T(v_j)^{-1} = U_j^{-1}Y'_j T_j^{-1} Y^{\alpha_j^\vee - \pi'_j} U_j \\ &= U_j^{-1} T_j U_j \end{aligned}$$

by (3.2.6), since $<\pi'_j, \alpha_j> = 1$. The proof of (3.3.1) is now complete. □

3.4 The double braid group

We have seen in the previous section that the braid group $\mathfrak{B}$ is generated by its subgroups $\mathfrak{B}_0$ and $Y^{L'}$, subject to the relations (3.2.4). We shall now iterate this construction. For this purpose, let R', L, and $\Lambda = L \oplus \mathbb{Z}c_0$ be as defined in §1.4, and for each $\alpha \in R$ let $\alpha' (= \alpha$ or $\alpha^\vee)$ be the corresponding element of R'. Let

$$X^\Lambda = \{X^f : f \in \Lambda\}$$

be a multiplicative group isomorphic to Λ, so that $X^f X^g = X^{f+g}$ and $(X^f)^{-1} = X^{-f}$ for $f, g \in \Lambda$.

The *double braid group* $\tilde{\mathfrak{B}}$ is the group generated by $\mathfrak{B}$ and X^Λ subject to the relations

$$T_i X^f T_i^\varepsilon = X^{s_i f}$$

for all $i \in I$ and $f \in \Lambda$ such that $<f, \alpha'_i> = 1$ or 0, where $\varepsilon = +1$ or -1 according as $<f, \alpha'_i> = 1$ or 0; and

$$U_j X^f U_j^{-1} = X^{u_j f}$$

for all $j \in J$ and $f \in \Lambda$. (As in §1.4, the elements of Λ are regarded as affine-linear functions on V.)

Let $q_0 = X^{c_0}$ and let $q = X^c = q_0^e$, where e is given by (1.4.5). The relations above show that q_0 commutes with each T_i and each U_j, and hence that q_0 is central in $\tilde{\mathfrak{B}}$. Also let

$$X^L = \{X^\lambda : \lambda \in L\}.$$

Then $\tilde{\mathfrak{B}}$ is generated by the groups $\mathfrak{B}_0$, X^L, $Y^{L'}$ and a central element q_0, subject to the following relations (3.4.1)–(3.4.5):

(3.4.1) $$T_i^\varepsilon Y^{-\lambda'} T_i = Y^{-s_i\lambda'}$$

for $i \in I, i \neq 0$ and $\lambda' \in L'$ such that either $<\lambda', \alpha_i> = 1$ and $\varepsilon = 1$, or $<\lambda', \alpha_i> = 0$ and $\varepsilon = -1$;

(3.4.2) $$T_i X^\lambda T_i^\varepsilon = X^{s_i\lambda}$$

for $i \in I, i \neq 0$ and $\lambda \in L$ such that either $<\lambda, \alpha_i'> = 1$ and $\varepsilon = 1$, or $<\lambda, \alpha_i'> = 0$ and $\varepsilon = -1$;

(3.4.3) $$T_0 X^\lambda T_0 = q^{-1} X^{s_\varphi \lambda}$$

where $\lambda \in L$ is such that $<\lambda, \varphi'> = -1$;

(3.4.4) $$T_0 X^\lambda = X^\lambda T_0$$

where $\lambda \in L$ is such that $<\lambda, \varphi'> = 0$;

(3.4.5) $$U_j X^\lambda U_j^{-1} = q^{-<\lambda, v_j \pi_j'>} X^{v_j^{-1}\lambda}$$

for $\lambda \in L$ and $j \in J$.

Here T_0 and U_j are defined by

(3.4.6) $$T_0 = Y^{\varphi^\vee} T(s_\varphi)^{-1},$$

(3.4.7) $$U_j = Y_j' T(v_j)^{-1},$$

where $Y_j' = Y^{\pi_j'}$.

If $J = \{0\}$, the relations (3.4.5) are absent. If $J \neq \{0\}$, the relations (3.4.3) and (3.4.4) are consequences of (3.4.2) and (3.4.5). For if $j \in J$ and $j \neq 0$, we have $T_0 = U_j^{-1} T_j U_j$ by (3.1.5), and therefore (3.4.3) and (3.4.4) come from (3.4.2) by conjugating with U_j^{-1} and using (3.4.5).

We observe next that (3.4.2) is obtained from (3.4.1) by replacing $Y^{\lambda'}$ by $X^{-\lambda}$ and reversing the order of multiplication. It follows that the results of §3.3, being consequences of (3.4.1) and the braid relations not involving T_0, have their counterparts, involving $\mathfrak{B}_0$ and the X^λ, in $\tilde{\mathfrak{B}}$. Thus, corresponding to (3.3.2) and (3.3.5) we have respectively

(3.4.8) *Let $\lambda \in L$ and (as in §2.4) let $v(\lambda)$ be the shortest element in W_0 such that $v(\lambda)\lambda$ is antidominant. Let $w \in W_0$ be such that $<\lambda, \alpha'> = 0$ or 1 for all $\alpha \in S(w)$. Then*

$$X^{w\lambda}T(wv(\lambda)^{-1})T(v(\lambda)^{-1})^{-1}X^{-\lambda} = T(w).$$

(3.4.9) *Let ψ' be the highest root of R'. If $w \in W_0$ is such that $\psi' = w^{-1}\alpha'_i$, where $i \neq 0$, then*

$$T(s_\psi)^{-1}X^{-\psi'^\vee} = T(w)T_i^{-1}X^{-a_i}T(w)^{-1}.$$

Let π_i $(i \in I, i \neq 0)$ be the fundamental coweights for R', defined by $<\pi_i, \alpha'_j> = \delta_{ij}$. Also define $\pi_0 = 0$. Let m'_i $(i \in I)$ be the integers attached to the nodes of the Dynkin diagram of $S(R')$, as in §1.3. As in §2.5, define a subset J' of I by

(3.4.10) *$k \in J'$ if and only if $\pi_k \in L$ and $m'_k = 1$.*

Let $X_k = X^{\pi_k}$ for $k \in J'$, and recall that $Y'_j = Y^{\pi'_j}$ for $j \in J$. Then we have the commutator formula

$$X_k^{-1}Y_j'^{-1}X_kY'_j = q^rT\left(w_k^{-1}\right)T\left(v_jw_k^{-1}\right)^{-1}T(v_j) \tag{3.4.11}$$

where $v_j = v(\pi'_j)$, $w_k = v(\pi_k)$, and $r = <\pi'_j, \pi_k>$.

Proof We have

$$\begin{aligned} Y_j'^{-1}X_kY'_j &= T(v_j)^{-1}U_j^{-1}X_kU_jT(v_j) \\ &= q^rT(v_j)^{-1}X^{v_j\pi_k}T(v_j) \quad \text{by (3.4.5)} \\ &= q^rX_kT\left(w_k^{-1}\right)T\left(v_jw_k^{-1}\right)^{-1}T(v_j) \end{aligned}$$

by (3.4.8) with $w = v_j$ and $\lambda = \pi_k$. □

3.5 Duality

Let $\tilde{\mathfrak{B}}'$ be the group obtained from $\tilde{\mathfrak{B}}$ by interchanging L and L'.

(3.5.1) *There is an anti-isomorphism ω of $\tilde{\mathfrak{B}}'$ onto $\tilde{\mathfrak{B}}$ in which $X^{\lambda'}$ $(\lambda' \in L')$, Y^λ $(\lambda \in L)$, T_i $(i \neq 0)$ and q_0 are mapped respectively to $Y^{-\lambda'}$, $X^{-\lambda}$, T_i and q_0.*

Let $\psi \in R$ be such that ψ' is the highest root of R'. Thus $\psi = \varphi$ if $R' = R$, and ψ is the highest short root of R if $R' = R^\vee \neq R$. In $\tilde{\mathfrak{B}}'$, T_0 is replaced by

$$Y^{\psi'^\vee} T(s_\psi)^{-1}$$

and U_j $(j \in J)$ by

$$Y^{\pi_k} T(w_k)^{-1}$$

where $k \in J'$ and $w_k = v(\pi_k)$. The images of these elements under ω are respectively

(3.5.2) $$T_0^* = T(s_\psi)^{-1} X^{-\psi'^\vee},$$

(3.5.3) $$V_k = T\left(w_k^{-1}\right)^{-1} X_k^{-1}.$$

Hence to prove (3.5.1) we have to show that in $\tilde{\mathfrak{B}}$ we have

(3.5.4) $$T_0^* Y^{\lambda'} T_0^* = q^{-1} Y^{s_\psi \lambda'}$$

for $\lambda' \in L'$ such that $<\lambda', \psi> = 1$;

(3.5.5) $$T_0^* Y^{\lambda'} = Y^{\lambda'} T_0^*$$

for $\lambda' \in L'$ such that $<\lambda', \psi> = 0$; and

(3.5.6) $$V_k Y^{\lambda'} V_k^{-1} = q^{-<\lambda', \pi_k>} Y^{w_k \lambda'}$$

for $\lambda' \in L'$ and $k \in J'$.

We remark that, as in the proof of (3.3.1), the defining relations (3.4.2) imply that T_0^* satisfies the appropriate braid relations (for the affine Weyl group of type R') and that

(3.5.7) $$V_j V_k = V_{j+k}$$

for $j, k \in J'$, and

(3.5.8) $$V_k^{-1} T_i V_k = \begin{cases} T_{i+k} & \text{if } i + k \neq 0, \\ T_0^* & \text{if } i + k = 0 \end{cases}$$

for $i \in I, i \neq 0$ and $k \in J'$.

Finally, it follows from the commutator formula (3.4.11) that

(3.5.9) $$V_k Y_j'^{-1} V_k^{-1} = q^r Y^{-w_k \pi_j'}$$

for $j \in J$ and $k \in J'$, where $r = <\pi'_j, \pi_k>$, i.e. that (3.5.6) is true for $\lambda' = -\pi'_j$. For

$$\begin{aligned} V_k Y_j'^{-1} V_k^{-1} &= T\left(w_k^{-1}\right)^{-1} X_k^{-1} Y_j'^{-1} X_k T\left(w_k^{-1}\right) \\ &= q^r T\left(v_j w_k^{-1}\right)^{-1} T(v_j) Y_j'^{-1} T\left(w_k^{-1}\right) \\ &= q^r Y^{-w_k \pi'_j} \end{aligned}$$

by (3.3.2) with $\lambda' = \pi'_j$ and $w = w_k^{-1}$. □

In §3.6 we shall prove (3.5.1) in the case $R' = R$, and in §3.7 in the case $R' \neq R$.

3.6 The case $R' = R$

In this section we shall assume that $R' = R$, so that $\psi = \varphi$. Then either $L = L' = P^\vee$ (1.4.2) or $L = L' = Q^\vee$ and R is of type C_n (1.4.3).

(a) Proof of (3.5.4) and (3.5.5)

When $J' \neq \{0\}$, (3.5.4) and (3.5.5) are consequences of (3.5.6) and the defining relations (3.4.1), since T_0^* can be conjugated into T_k, where $k \in J'$ and $k \neq 0$, by use of (3.5.8). Hence we may assume that $J = J' = \{0\}$, so that $R = R'$ is of type E_8, F_4, G_2 or C_n. Assume for the present that R is *not* of type C_n and (as in §1.4) that $|\varphi|^2 = 2$, so that

$$T_0^* = T(s_\varphi)^{-1} X^{-\varphi} = Y^{-\varphi} T_0 X^{-\varphi}.$$

The Dynkin diagrams in §1.3 show that in each case there is a unique long simple root α_1 such that $<\varphi, \alpha_1> \neq 0$ and $<\varphi, \alpha_i> = 0$ for all $i \neq 0, 1$. Hence it is enough to show that

(3.6.1) $$T_0^* Y^{\alpha_1} T_0^* = q^{-1} Y^{\alpha_1 - \varphi},$$

(3.6.2) $$T_0^* Y^{\alpha_i^\vee} = Y^{\alpha_i^\vee} T_0^*$$

for all $i \neq 0, 1$.

Proof of (3.6.1). From (3.4.1) and (3.4.2) we have

(a) $$T_1 X^\varphi T_1 = X^{\varphi - \alpha_1},$$

(b) $$T_0 X^{-\alpha_1} T_0 = q^{-1} X^{\varphi - \alpha_1},$$

(c) $$T_1 Y^{-\varphi} T_1 = Y^{-\varphi + \alpha_1}.$$

Hence

$$\begin{aligned}
T_0^* Y^{\alpha_1} T_0^* &= T_0^* Y^{\alpha_1-\varphi} T_0 X^{-\varphi} \\
&= T_0^* T_1 Y^{-\varphi} T_1 T_0 X^{-\varphi} \\
&= T_0^* T_1 T_0^* X^{\varphi} T_0^{-1} T_1 T_0 X^{-\varphi} \\
&= T_1 T_0^* T_1 X^{\varphi} T_1 T_0 T_1^{-1} X^{\varphi}
\end{aligned}$$

by use of (c) and the braid relations. This last expression is equal to $q^{-1} Y^{\alpha_1-\varphi}$ by successive application of (a), (b), (a), and (c). □

Proof of (3.6.2). Let D be the Dynkin diagram of $S = S(R)$, with vertex set I. Since D is a tree in the cases (E_8, F_4, G_2) under consideration, there is a unique path in D from 0 to any other vertex i. We proceed by induction on the length d of this path. The induction starts at $d = 2$, where we have $<\varphi, \alpha_i> = 0$, $<\varphi, \alpha_1> = 1$ and $<\alpha_1, \alpha_i> = -1$. It follows that T_0^* and T_i commute, and therefore

$$T_i T_0^* Y^{\alpha_1} T_0^* T_i = T_0^* T_i Y^{\alpha_1} T_i T_0^*.$$

By evaluating either side by use of (3.6.1) and (3.4.1), we see that T_0^* commutes with $Y^{\alpha_i^\vee}$.

Now let $d > 2$ and let $j \in I$ be the first vertex encountered on the path from i to 0 in D. We have $<\alpha_i, \alpha_j^\vee> = -1$, since either α_i and α_j are roots of the same length, or α_i is short and α_j is long. Hence by (3.4.2) we have

(1) $$T_i Y^{\alpha_i^\vee} T_i = Y^{\alpha_i^\vee + \alpha_j^\vee}.$$

Since T_0^* commutes with $Y_j^{\alpha^\vee}$ by the inductive hypothesis, and with T_i by the braid relations, it follows from (1) that T_0^* commutes with $Y^{\alpha_i^\vee}$.

There remains the case where $R = R'$ is of type C_n, and $L = L' = Q^\vee$. In this case the relations (3.5.4) are absent. In the notation of (1.3.4), we have to show that T_0^* commutes with Y^{ε_i} for $2 \leq i \leq n$. From (3.4.1) we have

(1) $$Y^{\varepsilon_{i+1}} = T_i^{-1} Y^{\varepsilon_i} T_i^{-1} \qquad (1 \leq i \leq n-1).$$

Now

$$t(\varepsilon_1) = s_0 s_1 \cdots s_n \cdots s_2 s_1$$

is a reduced expression, so that

(2) $$Y^{\varepsilon_1} = T_0 T_1 \cdots T_n \cdots T_2 T_1.$$

From (1) and (2) we have

$$Y^{\varepsilon_i} = T_{i-1}^{-1} \cdots T_1^{-1} T_0 T_1 \cdots T_n \cdots T_{i+1} T_i.$$

Since T_0^* commutes with $T_2, T_3, \ldots, T_n$, it is enough to show that T_0^* commutes with $T_1^{-1} T_0 T_1$, or equivalently that T_0 commutes with $T_1 T_0^* T_1^{-1}$. But $T_0^* = Y^{-\varepsilon_1} T_0 X^{-\varepsilon_1}$, hence

$$\begin{aligned}\left(T_1 T_0^* T_1^{-1}\right)^{-1} &= T_1 X^{\varepsilon_1} T_0^{-1} Y^{\varepsilon_1} T_1^{-1} \\ &= T_1 X^{\varepsilon_1} T_1 T_2 \cdots T_n \cdots T_3 T_2 \\ &= X^{\varepsilon_2} T_2 T_3 \cdots T_n \cdots T_3 T_2\end{aligned}$$

by (2) above and (3.4.2). Since $X^{\varepsilon_2}, T_2, T_3, \ldots, T_n$ all commute with T_0, so also does $T_1 T_0^* T_1^{-1}$. □

(b) Proof of (3.5.6)

Since L is generated by the π_j $(j \in J)$ and the coroots $\alpha_i^\vee$ $(i \in I, i \neq 0)$, it is enough by (3.5.9) to show that

(3.6.3) $$V_k Y^{\alpha_i^\vee} V_k^{-1} = q^{-<\pi_k, \alpha_i^\vee>} Y^{v_k \alpha_i^\vee}$$

Suppose first that $i = k$. By (3.3.7) we have

$$Y^{\alpha_k^\vee} = T_k T(v_k)^{-1} T_0 T(v_k)$$

and

$$V_k = V_{-k}^{-1} = X_{-k} T(v_k).$$

Since $V_k T_k V_k^{-1} = T_0^*$ by (3.5.8), it follows that

$$\begin{aligned}V_k Y^{\alpha_k^\vee} V_k^{-1} &= T_0^* X_{-k} T_0 X_{-k}^{-1} = q^{-1} T_0^* X^{\varphi} T^{-1} T_0 \quad \text{by (3.4.3)} \\ &= q^{-1} Y^{-\varphi}\end{aligned}$$

which proves (3.6.3) in this case, since $v_k \alpha_k^\vee = \alpha_0 = -\varphi$.

Now suppose that $i \neq 0, k$. As in the proof of (3.6.2) we proceed by induction on the length of a shortest path from i to 0 in the Dynkin diagram D of $S(R)$. Let j be the first vertex encountered on this path. We have $<\alpha_i, \alpha_j^\vee> = -1$ (the only exception to this statement is when R is of type C_n and α_i is the long simple root; but in that case $i = k$, which is excluded). By (3.5.8) we have

$$V_k T_i Y^{\alpha_j^\vee} T_i V_k^{-1} = T_{i-k} V_k Y^{\alpha_j^\vee} V_k^{-1} T_{i-k}.$$

On evaluating either side by use of (3.4.1) and the inductive hypothesis, we obtain (3.6.3). □

3.7 The case $R' \neq R$

In this section $R' = R^\vee \neq R$, so that R has two root lengths, and is of one of the types B_n, C_n, F_4, G_2. As in §1.4 we shall assume that L (resp L') is the lattice of weights of R (resp. $R^\vee$). As before, it is enough to verify (3.5.4) and (3.5.5) when $J' = \{0\}$ (i.e., when R is of type F_4 or G_2) and (3.5.6) when R is of type B_n or C_n.

Let φ be the highest root and ψ the highest short root of R. We have $<\psi, \alpha^\vee> = 0$ or 1 for all $\alpha \in R^+$ except $\alpha = \psi$, so that $<\psi, \varphi^\vee> = 0$ or 1. Moreover,

$$\varphi^\vee = \sum m_i' \alpha_i^\vee$$

where each m_i' is a positive integer (they are the labels for the affine root system $S(R)^\vee$), and hence

$$<\psi, \varphi^\vee> = \sum m_i' <\psi, \alpha_i^\vee> > 0.$$

It follows that $<\psi, \varphi^\vee> = 1$, and hence that $s_\varphi \psi$ and $s_\psi \varphi$ are negative roots.

(3.7.1) $$T_0^* Y^{\varphi^\vee} T_0^* = q^{-1} Y^{\varphi^\vee - \psi^\vee}$$

(i.e., (3.5.4) is true for $\lambda = \varphi^\vee$).

Proof In our present situation we have

$$T_0^* = T(s_\psi)^{-1} X^{-\psi}.$$

From (3.3.6) with $w = s_\psi$ we obtain

$$T_0 = T(s_\varphi s_\psi) Y^{\psi^\vee - \varphi^\vee} T(s_\psi)^{-1},$$

and from (3.4.8) with $w = s_\varphi s_\psi$ and $\mu = \psi$ we have

$$T_0^* = T(s_\varphi)^{-1} X^{\psi - \varphi} T(s_\varphi s_\psi).$$

Hence

$$\begin{aligned} Y^{\varphi^\vee} T_0^* &= Y^{\varphi^\vee} T(s_\varphi)^{-1} X^{\psi - \varphi} T(s_\varphi s_\psi) \\ &= T_0 X^{\psi - \varphi} T_0 T(s_\psi) Y^{\varphi^\vee - \psi^\vee} \\ &= q^{-1} X^\psi T(s_\psi) Y^{\varphi^\vee - \psi^\vee} \end{aligned}$$

by (3.4.3), and therefore

$$T_0^* Y^{\varphi^\vee} T_0^* = q^{-1} Y^{\varphi^\vee - \psi^\vee}. \qquad \square$$

(a) Proof of (3.5.4) and (3.5.5)

Let $L'_\psi = \{\lambda' \in L' : <\lambda', \psi> = 0\}$. In view of (3.7.1) it is enough to prove that T_0^* commutes with $Y^{\lambda'}$ for $\lambda' \in L'_\psi$.
(i) When R is of type G_2, L'_ψ is generated by $\alpha_1^\vee = 2\varphi^\vee - \psi^\vee$ (1.3.13). From (3.7.1) we have

$$T_0^* Y^{\varphi^\vee} T_0^* Y^{\psi^\vee - \varphi^\vee} = q^{-1} = T_0^* Y^{\psi^\vee - \varphi^\vee} T_0^* Y^{\varphi^\vee}.$$

Hence T_0^* commutes with $Y^{2\varphi^\vee - \psi^\vee}$.
(ii) When R is of type F_4, L'_ψ is generated by $\alpha_1^\vee, \alpha_2^\vee$ and $\alpha_3^\vee$, in the notation of (1.3.11). Since by (3.4.1)

$$T_2 Y^{\alpha_1^\vee} T_2 = Y^{\alpha_1^\vee + \alpha_2^\vee}, \quad T_3 Y^{\alpha_2^\vee} T_3 = Y^{\alpha_2^\vee + \alpha_3^\vee}$$

and since T_0^* commutes with T_2 and T_3, it is enough to verify that T_0^* commutes with $Y^{\alpha_1^\vee}$. Let

$$\lambda = \varphi^\vee = \varepsilon_1 + \varepsilon_2, \quad \mu = \varphi^\vee - \psi^\vee = -\varepsilon_1 + \varepsilon_2, \quad \nu = s_4 \mu = -\varepsilon_3 - \varepsilon_4$$

in the notation of (1.3.11) Then by (3.7.1) we have

$$T_0^* Y^\lambda T_0^* = q^{-1} Y^\mu, \quad T_4 Y^\mu T_4 = Y^\nu$$

and therefore

$$\begin{aligned} T_0^* Y^\nu &= q T_0^* T_4 T_0^* Y^\lambda T_0^* T_4 = q T_4 T_0^* T_4 Y^\lambda T_0^* T_4 \\ &= q T_4 T_0^* Y^\lambda T_0^* T_4 T_0^* = Y^\nu T_0^* \end{aligned}$$

by use of the braid relations and the fact that T_4 commutes with Y^λ. Hence T_0^* commutes with Y^ν. Finally, we have

$$T_1 Y^\nu T_1 = Y^{\nu - \alpha_1^\vee},$$

and since T_0^* commutes with Y^ν and T_1, it follows that T_0^* commutes with $Y^{\alpha_1^\vee}$. □

(b) Proof of (3.5.6)

As remarked above, we need only consider the cases where R is of type B_n or C_n $(n \geq 2)$.
(i) When R is of type B_n we have $J' = \{0, n\}$ in the notation of the list in §1.3, and it is enough to verify (3.5.6) when $\lambda = -\varepsilon_i$ $(1 \leq i \leq n)$, i.e. that

(3.7.2) $$V_n Y^{-\varepsilon_i} V_n^{-1} = q^{1/2} Y^{\varepsilon_{n+1-i}} \qquad (1 \leq i \leq n).$$

When $i = 1$, this follows from (3.5.9), since $\pi_1 = \varepsilon_1$. Using $T_i Y^{-\varepsilon_i} T_i = Y^{-\varepsilon_{i+1}} (1 \leq i \leq n-1)$, (3.7.2) now follows by induction on i.

(ii) When R is of type C_n we have $J' = \{0, 1\}$ in the notation of §1.3. In view of (3.5.9), it is enough to show that

(3.7.3) $$V_1 Y^{\varepsilon_1} V_1^{-1} = q^{-1} Y^{\varepsilon_1},$$

(3.7.4) $$V_1 Y^{\varepsilon_i} V_1^{-1} = Y^{\varepsilon_i} \qquad (2 \leq i \leq n).$$

We have $\varphi^\vee = \varepsilon_1$, hence

$$T_0 X^{-\varepsilon_1} = Y^{\varepsilon_1} T(s_\varphi)^{-1} X^{-\varepsilon_1} = Y^{\varepsilon_1} V_1^{-1} = Y^{\varepsilon_1} V_1$$

Hence

$$(Y^{\varepsilon_1} V_1)^2 = (T_0 X^{-\varepsilon_1})^2 = q^{-1}$$

which proves (3.7.3). Finally, from (3.7.1) we have

$$T_0^* Y^{\varepsilon_1} T_0^* = q^{-1} Y^{-\varepsilon_2},$$

and since $V_1 T_0^* V_1^{-1} = T_1$, it follows from (3.7.3) that

$$V_1 Y^{-\varepsilon_2} V_1^{-1} = T_1 Y^{-\varepsilon_1} T_1 = Y^{-\varepsilon_2}.$$

(3.7.4) now follows by induction on i, since $T_i Y^{-\varepsilon_i} T_i = Y^{-\varepsilon_{i+1}}$. □

The proof of (3.5.1) is now complete.

Notes and references

The braid group (also called the Artin group) associated to an arbitrary Coxeter group was studied by Brieskorn and Saito in [B2], and in van der Lek's thesis [V1]. The double braid group was introduced by Cherednik [C1], and the duality theorem (3.5.1) stated (in the case that $\tilde{\mathfrak{B}}' = \tilde{\mathfrak{B}}$). The commutator formula (3.4.11) is also due to Cherednik (private communication).

4

The affine Hecke algebra

4.1 The Hecke algebra of W

We retain the notation of the previous chapters: $W = W(R, L')$ is an extended affine Weyl group, and $\mathfrak{B}$ is the braid group of W.

The objects to be studied in this and subsequent chapters will involve certain parameters q, τ_i, and rational functions in these parameters. It would be possible to regard these parameters abstractly as independent indeterminates over $\mathbb{Z}$, but we shall find it more convenient to regard them as real variables. So let q be a real number such that $0 < q < 1$, and let τ_i $(i \in I)$ be positive real numbers such that $\tau_i = \tau_j$ if s_i and s_j are conjugate in W. Let K be a subfield of $\mathbb{R}$ containing the τ_i and $q_0 = q^{1/e}$, where e is the integer defined in (1.4.5).

The *Hecke algebra* $\mathfrak{H}$ of W over K is the quotient of the group algebra $K\mathfrak{B}$ of the braid group $\mathfrak{B}$ by the ideal generated by the elements

$$(T_i - \tau_i)(T_i + \tau_i^{-1})$$

for $i \in I$. The image of T_i (resp. $T(w)$, U_j) in $\mathfrak{H}$ will be denoted by the same symbol T_i (resp. $T(w)$, U_j). Thus $\mathfrak{H}$ is generated over K by T_i $(i \in I)$ and U_j $(j \in J)$, subject to the relations (3.1.3)–(3.1.5), together with the *Hecke relations*

(4.1.1) $$(T_i - \tau_i)(T_i + \tau_i^{-1}) = 0 \qquad (i \in I)$$

or equivalently

(4.1.1′) $$T_i - \tau_i = T_i^{-1} - \tau_i^{-1}.$$

(4.1.2) *Let $i \in I$, $w \in W$. Then in $\mathfrak{H}$ we have*

$$T_i T(w) = T(s_i w) + \chi(w^{-1} a_i)(\tau_i - \tau_i^{-1}) T(w),$$
$$T(w) T_i = T(w s_i) + \chi(w a_i)(\tau_i - \tau_i^{-1}) T(w),$$

where χ is the characteristic function of S^-.

Proof If $w^{-1}a_i \in S^+$ then $T_i T(w) = T(s_i w)$ by (3.1.7). If $w^{-1}a_i \in S^-$ then

$$T(s_i w) = T_i^{-1} T(w) = \left(T_i - \tau_i + \tau_i^{-1}\right) T(w)$$

by (3.1.7) and (4.1.1′). This proves the first of the relations above, and the second is proved similarly. □

(4.1.3) *The elements $T(w)$, $w \in W$, form a K-basis of $\mathfrak{H}$.*

Proof Let $\mathfrak{H}_1$ denote the K-subspace of $\mathfrak{H}$ spanned by the $T(w)$, $w \in W$. From (4.1.2) it follows that $T_i \mathfrak{H}_1 \subset \mathfrak{H}_1$ for all $i \in I$, and since $U_j T(w) = T(u_j w) \in \mathfrak{H}_1$, we have $U_j \mathfrak{H}_1 \subset \mathfrak{H}_1$ for all $j \in J$. Hence $\mathfrak{H}\mathfrak{H}_1 \subset \mathfrak{H}_1$, and since $1 \in \mathfrak{H}_1$ it follows that $\mathfrak{H} = \mathfrak{H}_1$, i.e. the $T(w)$ span $\mathfrak{H}$ as a K-vector space.

To show that the $T(w)$ are linearly independent, we proceed as follows. Let KW be the group algebra of W over K, and for each $i \in I$ define K-linear maps $L_i, R_i : KW \to KW$ by

$$L_i w = s_i w + \chi(w^{-1}a_i)\left(\tau_i - \tau_i^{-1}\right) w,$$
$$R_i w = w s_i + \chi(w a_i)\left(\tau_i - \tau_i^{-1}\right) w$$

for all $w \in W$ (compare (4.1.2)). Also, for each $u \in \Omega$ define L_u, R_u by

$$L_u w = uw, \quad R_u w = wu.$$

(4.1.4) *Each L commutes with each R.*

Proof It is clear that L_u commutes with each R, and that R_u commutes with each L. It remains to verify that L_i and R_j commute ($i, j \in I$). From the definitions we calculate that

$$(L_i R_j - R_j L_i) w = (\chi(w a_j) - \chi(s_i w a_j))\left(\tau_j - \tau_j^{-1}\right) s_i w$$
$$+ (\chi(s_j w^{-1} a_i) - \chi(w^{-1} a_i))\left(\tau_i - \tau_i^{-1}\right) w s_j.$$

Suppose first that $s_i w \neq w s_j$. Then $w a_j \neq \pm a_i$ and therefore $\chi(w a_j) = \chi(s_i w a_j)$ and $\chi(w^{-1}a_i) = \chi(s_j w^{-1} a_i)$. Hence $(L_i R_j - R_j L_i) w = 0$ in this case.

Suppose now that $s_i w = w s_j$, so that $\tau_i = \tau_j$ and $w a_j = \varepsilon a_i$, where $\varepsilon = \pm 1$. Then

$$\chi(w a_j) - \chi(w^{-1} a_i) = \chi(\varepsilon a_i) - \chi(\varepsilon a_j) = 0$$

and

$$\chi(s_i w a_j) - \chi(s_j w^{-1} a_i) = \chi(-\varepsilon a_i) - \chi(-\varepsilon a_j) = 0,$$

so that $(L_i R_j - R_j L_i) w = 0$ in this case also. □

Next we have

(4.1.5) $$L_i^2 = (\tau_i - \tau_i^{-1})L_i + 1$$

by a straightforward calculation.

Now let $\mathfrak{H}'$ denote the K-subalgebra of End(KW) generated by the L's, and let $f : \mathfrak{H}' \to KW$ be the linear mapping defined by $f(h) = h(1)$ for $h \in \mathfrak{H}'$.

(4.1.6) *$f : \mathfrak{H}' \to KW$ is an isomorphism (of K-vector spaces).*

Proof Let $w \in W$ and let $w = us_{i_1} \cdots s_{i_p}$ be a reduced expression (so that $u \in \Omega$ and $p = l(w)$). Since $L_i w = s_i w$ if $l(w) < l(s_i w)$ it follows that

$$f(L_u L_{i_1} \cdots L_{i_p}) = us_{i_1} \cdots s_{i_p} = w$$

and hence that f is surjective.

Suppose now that $h \in \text{Ker}(f)$. Then $h(1) = 0$, and we shall show by induction on $l(w)$ that $h(w) = 0$ for all $w \in W$. Suppose first that $l(w) = 0$, i.e. that $w = u \in \Omega$. From (4.1.4), R_u commutes with h, so that $h(u) = h(R_u(1)) = R_u h(1) = 0$. Now let $l(w) = p > 0$ and choose $i \in I$ so that $l(ws_i) < p$. Since R_i commutes with h, we have

$$h(w) = h(R_i(ws_i)) = R_i h(ws_i) = 0$$

by the inductive hypothesis. Hence $h = 0$ and f is an isomorphism. □

We can now complete the proof of (4.1.3). From (4.1.6) it follows that $L(w) := f^{-1}(w)$ is well-defined for all $w \in W$, and $L(w) = L_u L_{i_1} \cdots L_{i_p}$ if $w = us_{i_1} \cdots s_{i_p}$ is a reduced expression. Hence $L(v)L(w) = L(vw)$ if $l(v) + l(w) = l(vw)$, i.e. the $L(w)$ satisfy the defining relations (3.1.1) of the braid group $\mathfrak{B}$. From (4.1.5) it follows that $\mathfrak{H}'$ is a homomorphic image of $\mathfrak{H}$, i.e. that there is a surjective K-algebra homomorphism $g : \mathfrak{H} \to \mathfrak{H}'$ such that $g(T(w)) = L(w)$ for all $w \in W$. Hence $f \circ g : \mathfrak{H} \to KW$ maps $T(w)$ to w for each $w \in W$, and therefore the $T(w)$ are linearly independent over K. □

4.2 Lusztig's relation

We introduce the following notation: let

(4.2.1) $$\boldsymbol{b}(x) = \boldsymbol{b}(t, u; x) = \frac{t - t^{-1} + (u - u^{-1})x}{1 - x^2},$$

(4.2.2)
$$c(x) = c(t,u;x) = \frac{tx - t^{-1}x^{-1} + u - u^{-1}}{x - x^{-1}}$$
$$= \frac{(1-tux)(1+tu^{-1}x)}{t(1-x^2)} = c(t^{-1},u^{-1};x^{-1}).$$

where t, u are nonzero real numbers, and x is an indeterminate. When $t = u$, (4.2.1) and (4.2.2) take the simpler forms

$$b(x) = \frac{t - t^{-1}}{1-x}, \quad c(x) = \frac{tx - t^{-1}}{x-1}.$$

(4.2.3) *We have*

(i) $c(x) = t - b(x) = t^{-1} + b(x^{-1})$,
(ii) $c(x) + c(x^{-1}) = t + t^{-1}$,
(iii) $b(x) + b(x^{-1}) = t - t^{-1}$,
(iv) $c(x)c(x^{-1}) = 1 + b(x)b(x^{-1})$.

Proof (i) is clear, and (ii), (iii) follow directly from (i). As to (iv), we have

$$\begin{aligned} c(x)c(x^{-1}) &= (t - b(x))(t - b(x^{-1})) \\ &= t^2 - t(t - t^{-1}) + b(x)b(x^{-1}) \\ &= 1 + b(x)b(x^{-1}) \end{aligned}$$

by use of (i) and (iii). □

The following relation, due to Lusztig [L1], is fundamental.

(4.2.4) *Let $\lambda' \in L'$, $i \in I_0$. Then*

$$Y^{\lambda'}T_i - T_iY^{s_i\lambda'} = b(\tau_i, \upsilon_i; Y^{-\alpha_i^\vee})(Y^{\lambda'} - Y^{s_i\lambda'})$$

where $\upsilon_i = \tau_i$ or τ_0 according as $<L', \alpha_i> = \mathbb{Z}$ or $2\mathbb{Z}$ (2.1.6).

Proof If this formula (for fixed $i \in I_0$) is true for λ' and for μ', then it is immediate that it is true for $\lambda' + \mu'$ and for $-\lambda'$. Hence it is enough to prove it for λ' belonging to a fixed set of generators of L'.

If $<L', \alpha_i> = \mathbb{Z}$ (resp. $2\mathbb{Z}$), there exists $\mu' \in L'$ such that $<\mu', \alpha_i> = 1$ (resp. 2), and L' is generated by this μ' and the $\lambda' \in L'$ such that $<\lambda', \alpha_i> = 0$.

If $<\lambda', \alpha_i> = 0$, (4.2.4) reduces to $Y^{\lambda'}T_i = T_iY^{\lambda'}$, i.e. to (3.2.5).

If $<\lambda', \alpha_i> = 1$, from (3.2.6) and (4.1.1′) we have

$$T_iY^{s_i\lambda'} = Y^{\lambda'}T_i^{-1} = Y^{\lambda'}(T_i - \tau_i + \tau_i^{-1}),$$

so that

$$Y^{\lambda'}T_i - T_iY^{s_i\lambda'} = \left(\tau_i - \tau_i^{-1}\right)Y^{\lambda'},$$

which establishes (4.2.4) in this case, since $s_i\lambda' = \lambda' - \alpha_i^\vee$.

Finally, suppose that $<L', \alpha_i> = 2\mathbb{Z}$. From above, it is enough to verify (4.2.4) when $\lambda' = \alpha_i^\vee$. By (2.1.6) α_i is a long root, hence $\alpha_i = w^{-1}\varphi$ for some $w \in W_0$, where as usual φ is the highest root of R. From (3.3.5) we have

$$T_0 = T(w)Y^{\alpha_i^\vee}T_i^{-1}T(w)^{-1},$$

and hence $Y^{\alpha_i^\vee}T_i^{-1}$ is conjugate to T_0. It follows that

$$Y^{\alpha_i^\vee}T_i^{-1} - T_iY^{-\alpha_i^\vee} = \tau_0 - \tau_0^{-1}$$

by (4.1.1′), and hence

$$\begin{aligned} Y^{\alpha_i^\vee}T_i - T_iY^{-\alpha_i^\vee} &= \left(\tau_0 - \tau_0^{-1}\right) + \left(\tau_i - \tau_i^{-1}\right)Y^{\alpha_i^\vee} \\ &= \boldsymbol{b}(\tau_i, \tau_0; Y^{-\alpha_i^\vee})(Y^{\alpha_i^\vee} - Y^{-\alpha_i^\vee}) \end{aligned}$$

which completes the proof. □

(4.2.5) The right-hand side of the formula (4.2.4) is a linear combination of the Y's. Explicitly, if $<\lambda', \alpha_i> = r > 0$ it is

$$\sum_{j=0}^{r-1} u_j Y^{\lambda' - j\alpha_i^\vee}$$

and if $<\lambda', \alpha_i> = -r < 0$ it is

$$-\sum_{j=1}^{r} u_j Y^{\lambda' + j\alpha_i^\vee}$$

where $u_j = \tau_i - \tau_i^{-1}$ if j is even, and $u_j = \upsilon_i - \upsilon_i^{-1}$ if j is odd.

(4.2.6) In view of (4.2.3)(i), the formula (4.2.4) can be written in the equivalent forms

$$\begin{gathered} (T_i - \tau_i)Y^{\lambda'} - Y^{s_i\lambda'}(T_i - \tau_i) = \boldsymbol{c}(\tau_i, \upsilon_i; Y^{-\alpha_i^\vee})(Y^{s_i\lambda'} - Y^{\lambda'}), \\ \left(T_i + \tau_i^{-1}\right)Y^{\lambda'} - Y^{s_i\lambda'}\left(T_i + \tau_i^{-1}\right) = \boldsymbol{c}(\tau_i, \upsilon_i; Y^{\alpha_i^\vee})(Y^{\lambda'} - Y^{s_i\lambda'}). \end{gathered}$$

(4.2.7) *The elements $T(w)Y^{\lambda'}$ (resp. the elements $Y^{\lambda'}T(w)$), where $\lambda' \in L'$ and $w \in W_0$, form a K-basis of $\mathfrak{H}$.*

Proof Suppose that there is a relation of linear dependence

$$\sum_{i=1}^{r} u_i T(w_i) Y^{\lambda'_i} = 0$$

with distinct pairs $(w_i, \lambda'_i) \in W_0 \times L'$, and nonzero coefficients u_i. By multiplying on the right by a suitable $Y^{\mu'}$, we may assume that each λ'_i is dominant, and then by (2.4.1) we have $l(w_i t(\lambda'_i)) = l(w_i) + l(t(\lambda'_i))$, so that the relation above takes the form

$$\sum_{i=1}^{r} u_i T(w_i t(\lambda'_i)) = 0$$

contradicting (4.1.3). Hence the $T(w)Y^{\lambda'}$ are linearly independent over K, and a similar argument shows that the same is true of the $Y^{\lambda'}T(w)$.

Now let $\mathfrak{H}_1$ (resp. $\mathfrak{H}_2$) be the vector subspace of $\mathfrak{H}$ spanned by the $T(w)Y^{\lambda'}$ (resp. by the $Y^{\lambda'}T(w)$). By (4.2.4) and induction on $l(w)$, we see that $Y^{\lambda'}T(w) \in \mathfrak{H}_1$ and $T(w)Y^{\lambda'} \in \mathfrak{H}_2$, for all $w \in W_0$ and $\lambda' \in L'$. Hence $\mathfrak{H}_1 = \mathfrak{H}_2$. Now $\mathfrak{H}_1$ is stable under left multiplication by each $T(w)$ and (since $\mathfrak{H}_1 = \mathfrak{H}_2$) also by each $Y^{\lambda'}$. But these elements generate $\mathfrak{B}$ (3.3.1) and therefore also generate $\mathfrak{H}$ as K-algebra. Hence $\mathfrak{H}\mathfrak{H}_1 \subset \mathfrak{H}_1$, and since $1 \in \mathfrak{H}_1$ it follows that $\mathfrak{H} = \mathfrak{H}_1 = \mathfrak{H}_2$. □

Let $A' = KL'$ be the group algebra of the lattice L' over the field K. For each $\lambda' \in L'$ we denote the corresponding element of A' by $e^{\lambda'}$, so that

$$e^{\lambda'} \cdot e^{\mu'} = e^{\lambda' + \mu'}, \quad (e^{\lambda'})^{-1} = e^{-\lambda'}, \quad e^0 = 1$$

for $\lambda', \mu' \in L'$, and the $e^{\lambda'}$ form a K-basis of A'. The finite Weyl group W_0 acts on L' and hence on A':

$$w(e^{\lambda'}) = e^{w\lambda'}$$

for $w \in W_0$ and $\lambda' \in L'$.

If $f \in A'$, say

$$f = \sum f_{\lambda'} e^{\lambda'}$$

with coefficients $f_{\lambda'} \in K$, almost all zero, let

$$f(Y) = \sum f_{\lambda'} Y^{\lambda'}.$$

By (4.2.7) the $Y^{\lambda'}$, $\lambda' \in L'$, are linearly independent over K and span a commutative K-subalgebra $A'(Y)$ of $\mathfrak{H}$, isomorphic to A'. In this notation we may restate (4.2.4) as follows: for $i \neq 0$ and $f \in A'$ we have

$$f(Y)T_i - T_i(s_i f)(Y) = \boldsymbol{b}(\tau_i, \upsilon_i; Y^{-\alpha_i^\vee})(f(Y) - (s_i f)(Y)), \tag{4.2.8}$$

and the right-hand side is an element of $A'(Y)$.

By replacing f by $s_i f$ in (4.2.8), we see that $T_i f(Y)$ is of the form

$$T_i f(Y) = (s_i f)(Y)T_i + g(Y)$$

for some $g \in A'$. By induction on $l(w)$, it follows that for each $w \in W_0$ and $f \in A'$, $T(w)f(Y)$ is of the form

$$T(w)f(Y) = \sum_{v \leq w} f_v(Y)T(v) \tag{4.2.9}$$

where $f_v \in A'$, and is particular $f_w = wf$.

Let $A'_0 = (A')^{W_0}$ be the subalgebra of W_0-invariants in A'.

(4.2.10) *The centre of* $\mathfrak{H}$ *is* $A'_0(Y)$.

Proof Let $z \in \mathfrak{H}$ be a central element, say

$$z = \sum_{w \in W_0} f_w(Y)T(w)$$

with $f_w \in A'$. Let $\lambda' \in L'$ be regular (i.e., $\lambda' \neq w\lambda'$ for all $w \neq 1$ in W_0). Since z commutes with $Y^{\lambda'}$ we have

$$\sum_{v \in W_0} Y^{\lambda'} f_v(Y)T(v) = \sum_{w \in W_0} f_w(Y)T(w)Y^{\lambda'}. \tag{1}$$

Now by (4.2.9), $T(w)Y^{\lambda'}$ is of the form

$$T(w)Y^{\lambda'} = \sum_{v \leq w} g_{vw}(Y)T(v) \tag{2}$$

with $g_{vw} \in A'$ for each $v \leq w$, and $g_{ww} = e^{w\lambda'}$. From (1) and (2) we have

$$\sum_{v \in W_0} Y^{\lambda'} f_v(Y)T(v) = \sum_{\substack{v, w \in W_0 \\ w \geq v}} g_{vw}(Y) f_w(Y)T(v)$$

and hence by (4.2.7)

$$e^{\lambda'} f_v = \sum_{w \geq v} g_{vw} f_w \tag{3}$$

for each $v \in W_0$.

The matrix $G = (g_{vw})$, with rows and columns indexed by W_0, is triangular relative to any total ordering of W_0 that extends the Bruhat order. Its eigenvalues are therefore its diagonal elements, namely $e^{w\lambda'}$ ($w \in W_0$). If f denotes the column vector $(f_v)_{v \in W_0}$, the equation (3) shows that f is an eigenvector of G for the eigenvector $e^{\lambda'}$. Since the eigenvalues of G are all distinct (because λ' is regular), f is up to a scalar multiple the only eigenvector of G for the eigenvalue $e^{\lambda'}$. It follows that $f_v = 0$ for all $v \neq 1$ in W_0, and hence $z = f_1(Y) \in A'(Y)$.

Since z commutes with T_i, it follows from (4.2.8) that

$$T_i(f_1(Y) - (s_i f_1)(Y)) = g(Y)$$

for some $g \in A'$. Hence by (4.2.7) we have $f_1 = s_i f_1$ for each $i \neq 0$, and therefore $z \in A'_0(Y)$.

Conversely, if $f \in A'_0$ it follows from (4.2.8) that $f(Y)$ commutes with T_i for each $i \neq 0$, and hence $f(Y)$ is central in $\mathfrak{H}$. □

4.3 The basic representation of $\mathfrak{H}$

Let $\mathfrak{H}_0$ be the K-subalgebra of $\mathfrak{H}$ spanned by the elements $T(w)$, $w \in W_0$ (so that $\mathfrak{H}_0$ is the Hecke algebra of W_0). From (4.2.7) we have

(4.3.1) $$\mathfrak{H} \cong A' \otimes_K \mathfrak{H}_0$$

as K-vector spaces, the isomorphism being $Y^{\lambda'}T(w) \mapsto e^{\lambda'} \otimes T(w)$ ($\lambda' \in L'$, $w \in W_0$).

If M is a left $\mathfrak{H}_0$-module, we may form the induced $\mathfrak{H}$-module

$$\operatorname{ind}_{\mathfrak{H}_0}^{\mathfrak{H}}(M) = \mathfrak{H} \otimes_{\mathfrak{H}_0} M \cong A' \otimes_K M$$

by (4.3.1), the isomorphism being

$$f(Y)T(w) \otimes x \mapsto f \otimes T(w)x$$

for $f \in A'$, $w \in W_0$ and $x \in M$. From (4.2.8) it follows that the action of $\mathfrak{H}_0$ on $A' \otimes_K M$ is given by

(4.3.2) $$T_i(f \otimes x) = s_i f \otimes T_i x + (f - s_i f)\boldsymbol{b}(\tau_i, \upsilon_i; e^{-\alpha_i^\vee}) \otimes x.$$

In particular, let us take M to be the 1-dimensional $\mathfrak{H}_0$-module Kx for which $T_i x = \tau_i x$ for each $i \in I_0$. Then $A' \otimes_K M$ may be identified with A' (namely $f \otimes x \mapsto f$) and from (4.3.2) the action of $\mathfrak{H}_0$ on A' is given by

$$T_i(f) = \tau_i s_i f + (f - s_i f)\boldsymbol{b}(\tau_i, \upsilon_i; e^{-\alpha_i^\vee}).$$

Hence

(4.3.3) *There is a representation β' of $\mathfrak{H}_0$ on A' such that*

$$\beta'(T_i) = \tau_i s_i + \boldsymbol{b}(\tau_i, \upsilon_i; X^{-\alpha_i^\vee})(1 - s_i)$$

for all $i \in I_0$, where $X^{-\alpha_i^\vee}$ is the operator of multiplication by $e^{-\alpha_i^\vee}$, and $\upsilon_i = \tau_i$ or τ_0 according as $<L', \alpha_i> = \mathbb{Z}$ or $2\mathbb{Z}$.

In other words, the linear operators $\beta'(T_i) : A' \to A'$ satisfy the braid relations (3.1.3) and the Hecke relations (4.1.1) that do not involve T_0. In fact, this representation is faithful (see below).

From now on we shall assume that the conventions of §1.4 are in force, so that we have affine root systems S and S', finite root systems R and R', and lattices L and L', defined by (1.4.1)–(1.4.3). As in §1.4 the elements $\mu \in L$ are to be regarded as linear functions on $V : \mu(x) = <\mu, x>$ for $x \in V$. If $w \in W$ we shall denote the effect of w on μ so regarded by $w \cdot \mu$. Thus if $w = t(\lambda')v$, where $\lambda' \in L'$ and $v \in W_0$, then

$$(w \cdot \mu)(x) = \mu(w^{-1}x) = <\mu, v^{-1}(x - \lambda')> = <v\mu, x> - <\lambda', v\mu>$$

so that

(4.3.4) $$w \cdot \mu = v\mu - <\lambda', v\mu>c$$

is an affine-linear function on V.

Let $A = KL$ be the group algebra of L over K, and for each $\mu \in L$ let e^μ denote the corresponding element of A. More generally, if $f = \mu + rc$ we define

(4.3.5) $$e^f = q^r e^\mu$$

(i.e. we define e^c to be q). For f as above, let $X^f : A \to A$ denote multiplication by e^f:

(4.3.6) $$X^f g = e^f g \qquad (g \in A).$$

The group W acts on A: if $w = t(\lambda')v$ as above then

(4.3.7) $$w(e^\mu) = e^{w \cdot \mu} = q^{-<\lambda', v\mu>} e^{v\mu}$$

by (4.3.4).

(4.3.8) *W acts faithfully on A.*

For if $w = t(\lambda')v$ fixes e^μ for each $\mu \in L$, then (4.3.7) shows that $v\mu = \mu$ and $<\lambda', v\mu> = 0$, so that $v = 1$ and $\lambda' = 0$. □

When S is of type $(C_n^\vee, C_n)$ (1.4.3), we shall require two extra parameters τ_0', τ_n'. For uniformity of notation we define

$$\tau_i' = \tau_i$$

for all $i \in I$ when S is reduced ((1.4.1), (1.4.2)) and when $i \neq 0, n$ in case (1.4.3).

Let

(4.3.9)
$$\begin{aligned} \boldsymbol{b}_i &= \boldsymbol{b}(\tau_i, \tau_i'; e^{a_i}), \\ \boldsymbol{c}_i &= \boldsymbol{c}(\tau_i, \tau_i'; e^{a_i}) \end{aligned}$$

and for $\varepsilon = \pm 1$ let $\boldsymbol{b}_i(X^\varepsilon)$ (resp. $\boldsymbol{c}_i(X^\varepsilon)$) denote the result of replacing e^{a_i} by $X^{\varepsilon a_i}$ in $\boldsymbol{b}_i$ (resp. $\boldsymbol{c}_i$).

(4.3.10) *There is a faithful representation β of $\mathfrak{H}$ on A such that*

$$\begin{aligned} \beta(T_i) &= \tau_i s_i + \boldsymbol{b}_i(X)(1 - s_i), \\ \beta(U_j) &= u_j, \end{aligned}$$

for all $i \in I$ and $j \in J$, where as above X^{a_i} is multiplication by e^{a_i}.

Proof We saw above that the operators $\beta'(T_i)$, $i \neq 0$, defined in (4.3.3) satisfy the braid relations and the Hecke relations not involving T_0. Now $\mathfrak{H}_0$ depends only on W_0 (and the parameters τ_i), not on the particular root system R with W_0 as Weyl group. We may therefore replace (R, L') in (4.3.3) by (R', L), and the basis (α_i) of R by the opposite basis $(-\alpha_i')$ of R'. It follows that for $i \neq 0$ the operators $\beta(T_i)$ satisfy the braid relations and the Hecke relations.

Now the fact that $\beta(T_i)$ and $\beta(T_j)$ (where $i, j \neq 0$ and $i \neq j$) satisfy the appropriate braid and Hecke relations is a statement about the root system of rank 2 generated by a_i and a_j. It follows from this remark that the braid and Hecke relations involving $\beta(T_0)$ will also be satisfied. Moreover, it is clear from the definitions of $\beta(T_i)$ and $\beta(U_j) = u_j$ that the relations (3.1.4) and (3.1.5) are satisfied. Hence β is indeed a representation of $\mathfrak{H}$, and it remains to show that it is faithful. This will follow from □

(4.3.11) *The linear operators $X^\mu\beta(T(w))$ (resp. $\beta(T(w))X^\mu$), where $\mu \in L$ and $w \in W$, are linearly independent over K.*

Proof Let $w \in W$ and let $w = u_j s_{i_1} \cdots s_{i_p}$ be a reduced expression. Then

$$\beta(T(w)) = u_j \beta(T_{i_1}) \cdots \beta(T_{i_p})$$

and it follows from the definition (4.3.10) of $\beta(T_i)$ that $\beta(T(w))$ is of the form

(1) $$\beta(T(w)) = \sum_{v \le w} f_{vw}(X)v$$

where $f_{vw} \in \Phi$, the field of fractions of the integral domain A, and $f_{vw}(X)$ is the operator of multiplication by f_{vw}. We have $f_{ww} \neq 0$ for each $w \in W$.

Now suppose that the operators $X^\mu \beta(T(w))$ on A are linearly dependent. Then there will be a relation of the form

(2) $$\sum_{w \in W} g_w(X)\beta(T(w)) = 0$$

with $g_w \in A$ not all zero (but only finitely many nonzero). From (1) and (2) we have

$$\sum_{\substack{v,w \in W \\ v \le w}} g_w(X) f_{vw}(X)v = 0.$$

Now the automorphisms $v \in W$ of A extend to automorphisms of the field Φ, and as such are linearly independent over Φ, since automorphisms of any field are linearly independent over that field. Hence it follows from (4.3.8) that

(3) $$\sum_{w \ge v} f_{vw} g_w = 0$$

for each $v \in W$. Now choose v to be a maximal element, for the Bruhat ordering, of the (finite) set of $w \in W$ such that $g_w \neq 0$. Then (3) reduces to $f_{vv} g_v = 0$, and since $f_{vv} \neq 0$ it follows that $g_v = 0$. This contradiction shows that the operators $X^\mu \beta(T(w))$ are linearly independent.

For the operators $\beta(T(w))X^\mu$, the proof is similar. □

In particular, taking $\mu = 0$ in (4.3.11), it follows that the operators $\beta(T(w))$ $(w \in W)$ are linearly independent over K. Hence the representation β is faithful, completing the proof of (4.3.10). □

This representation β is the *basic representation* of $\mathfrak{H}$.

In view of (4.3.10), we may identify each $h \in \mathfrak{H}$ with the linear operator $\beta(h)$ on A. Since by (4.3.10)

$$T_i = \tau_i s_i + \boldsymbol{b}_i(X)(1 - s_i)$$

for each $i \in I$, it follows from (4.3.3) that

(4.3.12) $$T_i - \tau_i = \boldsymbol{c}_i(X)(s_i - 1),$$

(4.3.13) $$T_i + \tau_i^{-1} = (s_i + 1)\boldsymbol{c}_i(X^{-1}),$$

(4.3.14) $$T_i^{\varepsilon} = \varepsilon \boldsymbol{b}_i(X^{\varepsilon}) + \boldsymbol{c}_i(X)s_i,$$

where $\varepsilon = \pm 1$.

From (4.3.14) it follows that

(4.3.15) $$T_i X^{\mu} - X^{s_i\mu} T_i = \boldsymbol{b}_i(X)(X^{\mu} - X^{s_i\mu})$$

for all $\mu \in L$. In particular, we have

(4.3.16) $$T_i X^{\mu} = X^{\mu} T_i$$

if $<\mu, \alpha_i'> = 0$, and

(4.3.17) $$T_i X^{\mu} = X^{s_i\mu}\left(T_i - \tau_i + \tau_i^{-1}\right) = X^{s_i\mu} T_i^{-1}$$

if $<\mu, \alpha_i'> = 1$ (which implies that $\tau_i' = \tau_i$). Thus the X^{μ} satisfy the relations (3.4.2)–(3.4.5) for the double braid group $\tilde{\mathfrak{B}}$.

Recall that $A_0' = (A')^{W_0}$, and likewise let $A_0 = A^{W_0}$.

(4.3.18) *Let $f \in A_0'$. Then $f(Y)$ maps A_0 into A_0.*

Proof By (4.2.10), $f(Y)$ commutes with T_i for each $i \in I$. Let $g \in A_0$ and let $h = f(Y)g$. By (4.3.12) we have $T_i g = \tau_i g$, and hence

$$T_i h = T_i f(Y) g = f(Y) T_i g = \tau_i h$$

for all $i \neq 0$. By (4.3.12) it follows that $s_i h = h$ for all $i \neq 0$, hence $h \in A_0$. □

In the case (1.4.3), let

(4.3.19) $$T_n' = X^{-a_n} T_n^{-1}, \quad T_0' = X^{-a_0} T_0^{-1}$$

(where $a_0 = -\varepsilon_1 + \frac{1}{2}c$, so that $X^{-a_0} = q^{-1/2} X^{\varepsilon_1}$). Then we have

(4.3.20) $$(T_n' - \tau_n')\left(T_n' + \tau_n'^{-1}\right) = (T_0' - \tau_0')\left(T_0' + \tau_0'^{-1}\right) = 0.$$

Proof We calculate

$$\begin{aligned} T_n' - T_n'^{-1} &= X^{-a_n} T_n^{-1} - T_n X^{a_n} \\ &= X^{-a_n}\left(T_n - \tau_n + \tau_n^{-1}\right) - T_n X^{a_n} \\ &= \boldsymbol{b}_n(X)(X^{-a_n} - X^{a_n}) - \left(\tau_n - \tau_n^{-1}\right) X^{-a_n} \\ &= \tau_n' - \tau_n'^{-1} \end{aligned}$$

by (4.3.15) and (4.2.1); and likewise

$$T_0' - T_0'^{-1} = \tau_0' - \tau_0'^{-1}. \qquad \square$$

(4.3.21) From (4.3.12) it follows that, for $\lambda \in L$ and $i \neq 0$,
(i) if $<\lambda, \alpha_i'> = r > 0$, then

$$T_i e^\lambda = \tau_i^{-1} e^{s_i\lambda} - \sum_{j=1}^{r-1} u_j e^{\lambda - j\alpha_i};$$

(ii) if $<\lambda, \alpha_i'> = -r < 0$, then

$$T_i e^\lambda = \tau_i e^{s_i\lambda} + \sum_{j=0}^{r-1} u_j e^{\lambda + j\alpha_i},$$

where (as in (4.2.5))

$$u_j = \begin{cases} \tau_i - \tau_i^{-1} & \text{if } j \text{ is even,} \\ \tau_i' - \tau_i'^{-1} & \text{if } j \text{ is odd.} \end{cases}$$

We shall make use of the following terminology. If $f \in A$ is of the form

$$f = \sum_{\mu \leq \lambda} u_\mu e^\mu$$

where the partial ordering is that defined in §2.7 (with L' replaced by L), we shall often write

$$f = u_\lambda e^\lambda + \text{ lower terms.}$$

With this terminology we have

(4.3.22) *Let $\lambda \in L, i \neq 0$. Then*

$$T_i^{-1} e^\lambda = \tau_i^\varepsilon e^{s_i\lambda} + \text{lower terms,}$$

where $\varepsilon = -1$ if $<\lambda, \alpha_i'> \geq 0$, and $\varepsilon = +1$ if $<\lambda, \alpha_i'> < 0$.

Proof Since $T_i^{-1} = T_i - \tau_i + \tau_i^{-1}$, it follows from (4.3.21) that

$$T_i^{-1} e^\lambda = \tau_i^{-1} e^{s_i\lambda} - \sum_{j=0}^{r-1} u_j e^{\lambda - j\alpha_i}$$

if $<\lambda, \alpha_i'> = r > 0$. In this case we have $s_i\lambda > \lambda$, by (2.7.9).

Next, if $<\lambda, \alpha_i'> = -r < 0$ we have, using (4.3.21) again,

$$T_i^{-1}e^{\lambda} = \tau_i e^{s_i\lambda} + \sum_{j=1}^{r-1} u_j e^{\lambda + j\alpha_i}$$

which gives the result in this case.

Finally, if $<\lambda, \alpha_i'> = 0$ then $s_i\lambda = \lambda$ and $T_i^{-1}e^{\lambda} = \tau_i^{-1}e^{\lambda}$. □

For the remainder of this section, we need to switch to additive notation. We shall write

$$\tau_i = q^{\kappa_i/2}$$

and for $\alpha \in R$ we define

$$\kappa_\alpha = \kappa_i \quad \text{if} \quad \alpha \in W_0\alpha_i.$$

With this notation we have

(4.3.23) *Let $w \in W_0$, $\lambda \in L$. Then*

$$T(w^{-1})^{-1}e^{\lambda} = q^{f(w,\lambda)}e^{w\lambda} + \text{lower terms},$$

where

$$f(w, \lambda) = \frac{1}{2}\sum_{\alpha \in R^+} \eta(-<\lambda, \alpha'>)\chi(w\alpha)\kappa_\alpha$$

and χ is the characteristic function of R^-, and η is given by (2.8.3).

Proof Let $w = s_{i_1} \cdots s_{i_p}$ be a reduced expression, so that

$$T(w^{-1})^{-1} = T_{i_1}^{-1} \cdots T_{i_p}^{-1}.$$

From (4.3.22) it follows that

$$T(w^{-1})^{-1}e^{\lambda} = \left(\prod_{r=1}^{p} \tau_{i_r}^{\varepsilon_r}\right) e^{w\lambda} + \text{lower terms},$$

where

$$\begin{aligned}\varepsilon_r &= \eta(-<s_{i_{r+1}} \cdots s_{i_p}\lambda, \alpha_{i_r}'>) \\ &= \eta(-<\lambda, \beta_r'>)\end{aligned}$$

and $\beta_r' = s_{i_p} \cdots s_{i_{r+1}}\alpha_{i_r}'$, so that $\beta_1', \ldots, \beta_p'$ are the roots $\alpha' \in R'^+$ such that $w\alpha' \in R'^-$. It follows that

$$\prod_{i=1}^{p} \tau_{i_r}^{\varepsilon_r} = q^{f(w,\lambda)}$$

where

$$f(w,\lambda) = \frac{1}{2}\sum_{r=1}^{p} \eta(-<\lambda, \beta_r'>)\kappa_{\beta_r}$$
$$= \frac{1}{2}\sum_{\alpha\in R^+} \eta(-<\lambda, \alpha'>)\chi(w\alpha)\kappa_\alpha. \qquad \square$$

If $w \in W_0$ and $w = s_{i_1} \cdots s_{i_p}$ is a reduced expression, as above, let

$$\tau_w = \tau_{i_1} \cdots \tau_{i_p}.$$

This is independent of the reduced expression, and we have

(4.3.24) $$\tau_w = q^{g(w)}$$

where

$$g(w) = \frac{1}{2}\sum_{\alpha\in R^+} \chi(w\alpha)\kappa_\alpha.$$

This follows from (4.3.23) when λ is antidominant (so that $<\lambda, \alpha'> \leq 0$ for all $\alpha \in R^+$).

In particular:

(4.3.25) *Let $\lambda \in L$. Then*

$$g(v(\lambda)) = \frac{1}{4}\sum_{\alpha\in R^+} (1 + \eta(<\lambda, \alpha'>)\kappa_\alpha.$$

For by (2.4.4) $v(\lambda)\alpha' \in R'^-$ if and only if $<\lambda, \alpha'> > 0$, so that $\chi(v(\lambda)\alpha') = \frac{1}{2}(1 + \eta(<\lambda, \alpha'>)$.

4.4 The basic representation, continued

We shall use the parameters τ_i, τ_i' to define a labelling k of S as follows. Define κ_i, κ_i' $(i\in I)$ by

(4.4.1) $$\tau_i = q^{\kappa_i/2}, \quad \tau_i' = q^{\kappa_i'/2}.$$

Recall (§1.3) that

$$S_1 = \left\{a \in S : \tfrac{1}{2}a \notin S\right\} = \bigcup_{i\in I} Wa_i$$

(so that $S_1 = S$ if S is reduced, and $S_1 = S(R)^\vee$ where R is of type C_n in the situation of (1.4.3)). Then for $a \in S_1$ we define

(4.4.2) $$k(a) = \tfrac{1}{2}(\kappa_i + \kappa_i'), \quad k(2a) = \tfrac{1}{2}(\kappa_i - \kappa_i')$$

if $a \in Wa_i$. Note that $k(2a) = 0$ if $2a \notin S$.

Thus if S is reduced we have $k(a) = \kappa_i$ for $a \in Wa_i$, and if S is of type $(C_n^\vee, C_n)$ the labels $k_1, \dots, k_5$ are given by

(4.4.3) $$(k_1, k_2, k_3, k_4, k_5) = \left(\tfrac{1}{2}(\kappa_n + \kappa_n'), \tfrac{1}{2}(\kappa_n - \kappa_n'), \tfrac{1}{2}(\kappa_0 + \kappa_0'), \tfrac{1}{2}(\kappa_0 - \kappa_0'), \kappa\right)$$

where $\kappa = \kappa_1 = \kappa_2 = \cdots = \kappa_{n-1}$. Passing to the dual labelling $(k_1', \dots, k_5')$ (1.5.1) corresponds to interchanging κ_0 and κ_n'.

For each $a \in S_1$, if $a = wa_i$ ($w \in W, i \in I$) we define

(4.4.4) $$\tau_a = \tau_i, \tau_a' = \tau_i'$$

and

(4.4.5) $$\begin{aligned} \boldsymbol{b}_a &= \boldsymbol{b}_{a,k} = \boldsymbol{b}(\tau_a, \tau_a'; e^a), \\ \boldsymbol{c}_a &= \boldsymbol{c}_{a,k} = \boldsymbol{c}(\tau_a, \tau_a'; e^a) \end{aligned}$$

so that $\boldsymbol{b}_a = w\boldsymbol{b}_i$ and $\boldsymbol{c}_a = w\boldsymbol{c}_i$. Also, for each $w \in W$, let

(4.4.6) $$\boldsymbol{c}(w) = \boldsymbol{c}_{S,k}(w) = \prod_{a \in S_1(w)} \boldsymbol{c}_{a,k}.$$

Let $A[\boldsymbol{c}]$ denote the K-subalgebra of the field of fractions of A generated by A and the $\boldsymbol{c}_a$, $a \in S$.

(4.4.7) *Let $u, v \in W$. Then (as operators on A)*

$$T(u)^{-1}T(v) = \sum_{w \le u^{-1}v} f_w(X)w$$

where $f_w \in A[\boldsymbol{c}]$, and in particular

$$f_{u^{-1}v} = \boldsymbol{c}_{S,k}(v^{-1}u).$$

Proof Let $u^{-1}v = u_j s_{i_1} \dots s_{i_p}$ be a reduced expression. From (3.1.9) we have

$$T(u)^{-1}T(v) = u_j T_{i_1}^{\varepsilon_1} \cdots T_{i_p}^{\varepsilon_p}$$

where each of $\varepsilon_1, \ldots, \varepsilon_p$ is ± 1. By (4.3.14) it follows that

$$T(u)^{-1}T(v) = u_j\big(\boldsymbol{c}_{i_1}(X)s_{i_1} + \varepsilon_1\boldsymbol{b}_{i_1}(X^{\varepsilon_1})\big)\cdots\big(\boldsymbol{c}_{i_p}(X)s_{i_p} + \varepsilon_p\boldsymbol{b}_{i_p}(X^{\varepsilon_p})\big)$$

which on expansion is of the stated form, with leading term

$$u_j\boldsymbol{c}_{i_1}(X)s_{i_1}\boldsymbol{c}_{i_2}(X)s_{i_2}\cdots s_{i_{p-1}}\boldsymbol{c}_{i_p}(X)s_{i_p}$$

so that

$$f_{u^{-1}v} = \boldsymbol{c}_{b_1}\boldsymbol{c}_{b_2}\cdots\boldsymbol{c}_{b_p}$$

where $b_r = u_j s_{i_1}\cdots s_{i_{r-1}}(a_{i_r})$ for $1 \le r \le p$. From (2.2.2) and (2.2.9) it follows that $\{b_1, \ldots, b_p\} = S_1(v^{-1}u)$. □

(4.4.8) *Let $\lambda' \in L'$. Then (as operators on A)*

(i) $Y^{\lambda'} = \boldsymbol{c}(u(\lambda')^{-1})(X)u(\lambda')T(v(\lambda')) + \sum_{\substack{w\in W\\ w(0)<\lambda'}} g_w(X)w,$

(ii) $Y^{-\lambda'} = T(v(\lambda'))^{-1}\boldsymbol{c}(u(\lambda'))(X)u(\lambda')^{-1} + \sum_{\substack{w\in W\\ w(0)<\lambda'}} g'_w(X)w^{-1},$

where $g_w, g'_w \in A[\boldsymbol{c}]$, and $u(\lambda')$, $v(\lambda')$ are as defined in §2.4.

Proof (i) Let $u(\lambda') = u_j s_{i_1}\cdots s_{i_p}$ be a reduced expression. From (3.2.10) we have

$$Y^{\lambda'} = u_j T_{i_1}^{\varepsilon_1}\cdots T_{i_p}^{\varepsilon_p}T(v(\lambda'))$$

where each exponent ε_r is ± 1, so that as in (4.4.7)

$$u_j T_{i_1}^{\varepsilon_1}\cdots T_{i_p}^{\varepsilon_p} = \sum_{w\le u(\lambda')} f_w(X)w$$

with $f_w \in A[\boldsymbol{c}]$, and leading term

$$f_{u(\lambda')}(X)u(\lambda') = \boldsymbol{c}(u(\lambda')^{-1})(X)u(\lambda').$$

Hence

$$Y^{\lambda'} = \boldsymbol{c}(u(\lambda')^{-1})(X)u(\lambda')T(v(\lambda')) + \sum_{\substack{w\in W\\ w<u(\lambda')}} f_w(X)wT(v(\lambda')).$$

Now $T(v(\lambda'))$ is of the form

$$T(v(\lambda')) = \sum_{v\le v(\lambda')} h_v(X)v$$

with $h_v \in A[\boldsymbol{c}]$. Hence

$$Y^{\lambda'} = \boldsymbol{c}(u(\lambda')^{-1})(X)u(\lambda')T(v(\lambda')) + \sum_{w \in W} g_w(X)w$$

summed over $w \in W$ of the form $w = w'v$, where $w' < u(\lambda')$ and $v \leq v(\lambda')$ (so that $v \in W_0$). For each such w we have $w(0) = w'(0) < \lambda'$ by (2.7.12).
(ii) We have

$$Y^{-\lambda'} = T(v(\lambda'))^{-1} T_{i_p}^{-\varepsilon_p} \cdots T_{i_1}^{-\varepsilon_1} u_j^{-1}$$

and

$$T_{i_p}^{-\varepsilon_p} \cdots T_{i_1}^{-\varepsilon_1} u_j^{-1} = \sum_{w \leq u(\lambda')} f'_w(X)w^{-1}$$

with $f'_w \in A[\boldsymbol{c}]$, and leading term

$$f'_{u(\lambda')}(X)u(\lambda')^{-1} = \boldsymbol{c}(u(\lambda'))(X)u(\lambda')^{-1}.$$

Hence

$$Y^{-\lambda'} = T(v(\lambda'))^{-1}\boldsymbol{c}(u(\lambda'))(X)u(\lambda')^{-1} + \sum_{w < u(\lambda')} T(v(\lambda'))^{-1} f'_w(X)w^{-1}.$$

Now $T(v(\lambda'))^{-1}$ is of the form

$$T(v(\lambda'))^{-1} = \sum_{v \leq v(\lambda')} h'_v(X)v^{-1}$$

with $h'_v \in A[\boldsymbol{c}]$, and therefore

$$Y^{-\lambda'} = T(v(\lambda'))^{-1}\boldsymbol{c}(u(\lambda'))(X)u(\lambda')^{-1} + \sum_{w \in W} g'_w(X)w^{-1}$$

summed over $w \in W$ of the form $w = w'v$, where $w' < u(\lambda')$ and $v \leq v(\lambda')$. For each such w we have $w(0) = w'(0) < \lambda'$, as before. □

In particular:

(4.4.9) *Let $\lambda' \in L'$ be antidominant (i.e., $w_0\lambda'$ dominant). Then*

$$Y^{\lambda'} = \boldsymbol{c}(t(-\lambda'))(X)t(\lambda') + \sum_{w(0) < \lambda'} g_w(X)w$$

with $g_w \in A[\boldsymbol{c}]$.

For in this case $v(\lambda') = 1$ and $u(\lambda') = t(\lambda')$. □

For $\lambda' \in L'$ antidominant, let $\Sigma(\lambda')$ be the smallest saturated subset of L' that contains λ', as in §2.6, and let

$$\Sigma^0(\lambda') = \Sigma(\lambda') - W_0\lambda'. \tag{4.4.10}$$

Also let

$$m_{\lambda'} = \sum_{\mu' \in W_0\lambda'} e^{\mu'}. \tag{4.4.11}$$

By (4.3.18), $m_{\lambda'}(Y)$ maps A_0 into A_0. Let $m_{\lambda'}(Y)_0$ denote the restriction of $m_{\lambda'}(Y)$ to A_0. Then

$$m_{\lambda'}(Y)_0 = \sum_{w \in W_0^{\lambda'}} (w\boldsymbol{c}(t(-\lambda')))(X)t(w\lambda') + \sum_{\mu' \in \Sigma^0(\lambda')} g_{\mu'}(X)t(\mu') \tag{4.4.12}$$

where $g_{\mu'} \in A[\boldsymbol{c}]$, and $W_0^{\lambda'}$ is a transversal of the isotropy group of λ' in W_0.

Proof Let $\mu' \in W_0\lambda'$. If $\mu' \neq \lambda'$ then $\mu' < \lambda'$ and therefore, by (4.4.8) (i), $t(\lambda')$ does not occur in $Y^{\mu'}$. Hence by (4.4.9) the only term in $m_{\lambda'}(Y)$ that contains $t(\lambda')$ is $\boldsymbol{c}(t(-\lambda'))(X)t(\lambda')$, and (4.4.12) therefore follows from (4.4.9). □

There are two cases in which (4.4.12) leads to an explicit formula. The first is when $w_0\lambda'$ is a minuscule fundamental weight (i.e., $\lambda' = w_0\pi'_j$ for some $j \in J$, $j \neq 0$), in which case $\Sigma^0(\lambda')$ is empty. The second is when $\lambda' = -\varphi^\vee$, in which case $\Sigma^0(\lambda') = \{0\}$. These two cases provide precisely the operators used in [M5] to construct orthogonal polynomials.

4.5 The basic representation, continued

We shall regard each element f of A (or of A') as a function on V, as follows: if $x \in V$ and

$$f = \sum f_\lambda e^\lambda$$

with coefficients $f_\lambda \in K$, we define

$$f(x) = \sum f_\lambda q^{<\lambda, x>}. \tag{4.5.1}$$

Likewise, if h is an element of the field of fractions of A (or A'), say $h = f/g$, we define $h(x) = f(x)/g(x)$ at all points $x \in V$ where $g(x) \neq 0$. Thus for example $\boldsymbol{c}_i(x)$ is well-defined at all points $x \in V$ such that $a_i(x) \neq 0$.

We shall assume until further notice that the labels $k(a)$, $a \in S$, are nonzero. Recall (2.8.1) that for $\lambda' \in L'$,

$$r'_k(\lambda') = u(\lambda')(-\rho'_k).$$

(4.5.2) *Let $\lambda' \in L', i \in I$. If $\lambda' = s_i\lambda'$, then $\boldsymbol{c}_i(r'_k(\lambda')) = 0$.*

Proof From (2.8.4) (iii) it follows that

$$a_i(r'_k(\lambda')) + k(\alpha_i'^{\vee}) = 0$$

and therefore $\boldsymbol{c}_i(r'_k(\lambda'))$ is well-defined.

If $<L', \alpha_i> = \mathbb{Z}$, then $k(\alpha_i'^{\vee}) = \kappa_i$ (4.4.2) and

$$\boldsymbol{c}_i = q^{-\kappa_i/2}(1 - q^{\kappa_i}e^{a_i})/(1 - e^{a_i}).$$

Hence $\boldsymbol{c}_i(r'_k(\lambda')) = 0$.

If $<L', \alpha_i> = 2\mathbb{Z}$, we are in the situation of (1.4.3), so that $i = 0$ or n and $L = L' = \mathbb{Z}^n$. If $i = n$ we have $k(\alpha_n'^{\vee}) = k_1$, and

$$\boldsymbol{c}_n = \frac{(1 - q^{k_1}e^{a_n})(1 + q^{k_2}e^{a_n})}{q^{(k_1+k_2)/2}(1 - e^{2a_n})}$$

so that again $\boldsymbol{c}_n(r'_k(\lambda')) = 0$. Finally, if $i = 0$ we have $\lambda' \neq s_0\lambda'$ for all $\lambda' \in L'$ (since $<\lambda', \alpha_0>$ is an even integer), so this case cannot arise. □

(4.5.3) *Let $h \in \mathfrak{H}$, say*

$$h = \sum_{w \in W} h_w(X)w^{-1}$$

as an operator on A, where $h_w \in A[\boldsymbol{c}]$. If $\lambda' \in L'$ is such that $h_w(r'_k(\lambda')) \neq 0$ for some $w \in W$, then $w(r'_k(\lambda')) = r'_k(w\lambda')$.

Proof Since the $T(w)$, $w \in W$, form a K-basis of $\mathfrak{H}$, we have

(1) $$h = \sum_{v \in W} a_v T(v)$$

with coefficients $a_v \in K$. For each $v \in W$, by (4.4.7) we can write

(2) $$T(v) = \sum_{w} f_{vw}(X)w^{-1}$$

with $f_{vw} \in A[\boldsymbol{c}]$. From (1) and (2) we have

$$h_w = \sum_{v} a_v f_{vw}$$

for each $w \in W$, by (4.3.8). Hence if $h_w(r'_k(\lambda')) \neq 0$ we must have $f_{vw}(r'_k(\lambda')) \neq 0$ for some $v \in W$, and therefore we may assume that $h = T(v)$. We proceed by induction on $l(v)$.

If $l(v) = 0$, then $v \in \Omega$ and $f_{vw} = 1$ if $w = v^{-1}$, and $f_{vw} = 0$ otherwise. By (2.8.4) (i) we have $v^{-1}(r'_k(\lambda')) = r'_k(v^{-1}\lambda')$, which proves the result in this case.

If $l(v) > 0$, let $v = v's_i$ where $l(v') = l(v) - 1$. Then

$$T(v) = T(v')T_i = \left(\sum_w f_{v'w}(X)w^{-1}\right)(\boldsymbol{c}_i(X)s_i + b_i(X))$$

by (4.3.14), so that

$$f_{vw}(X) = f_{v',s_iw}(X)(w^{-1}s_i\boldsymbol{c}_i)(X) + f_{v'w}(X)(w^{-1}\boldsymbol{b}_i)(X)$$

and therefore

$$f_{vw}(r'_k(\lambda')) = f_{v',s_iw}(r'_k(\lambda'))\boldsymbol{c}_i(s_iw(r'_k(\lambda'))) + f_{v'w}(r'_k(\lambda'))\boldsymbol{b}_i(w(r'_k(\lambda'))).$$

Now suppose that $f_{vw}(r'_k(\lambda')) \neq 0$. Then either $f_{v'w}(r'_k(\lambda')) \neq 0$, in which case $w(r'_k(\lambda')) = r'_k(w\lambda')$ by the inductive hypothesis; or $f_{v',s_iw}(r'_k(\lambda')) \neq 0$ and $\boldsymbol{c}_i(s_iw(r'_k(\lambda'))) \neq 0$.

Let $\mu' = s_iw\lambda'$. Then we have

(3) $$s_iw(r'_k(\lambda')) = r'_k(\mu')$$

by the inductive hypothesis, and hence $\boldsymbol{c}_i(r'_k(\mu')) \neq 0$, so that by (4.5.2) $\mu' \neq s_i\mu'$ and therefore

(4) $$r'_k(s_i\mu') = s_i(r'_k(\lambda'))$$

by (2.8.4) (ii). From (3) and (4) it follows that $w(r'_k(\lambda')) = r'_k(w\lambda')$ as required. □

Let $w \in W_0$ and let $w = s_{i_1} \cdots s_{i_p}$ be a reduced expression. Define

(4.5.4) $$\tau_w = \tau_{i_1} \cdots \tau_{i_p}$$

which is independent of the reduced expression chosen. Since $T_i(1_A) = \tau_i 1_A$ for each i, where 1_A is the identity element of A, it follows that

(4.5.5) $$T(w)(1_A) = \tau_w 1_A.$$

(4.5.6) *Let $w \in W_0$ and let*

$$T(w) = \sum_v f_v(X)v^{-1}$$

with $f_v \in A[\mathbf{c}]$. Then $f_v(-\rho'_k) = 0$ if $v \neq 1$, and $f_1(-\rho'_k) = \tau_w$.

Proof We shall apply (4.5.3) with $\lambda' = 0$, so that $r'_k(\lambda') = -\rho'_k$. If $f_v(-\rho'_k) \neq 0$, then

$$v(-\rho'_k) = r'_k(v(0)) = r'_k(0) = -\rho'_k$$

whence $v = 1$ by (1.5.5).

Now evaluate both sides at 1_A. By (4.5.5) we obtain

$$\tau_w = \sum_v f_v.$$

Evaluating at $-\rho'_k$ now gives $f_1(-\rho'_k) = \tau_w$. □

(4.5.7) *Let $f \in A'_0$ and let $f(Y^{-1})_0$ denote the restriction of $f(Y^{-1})$ to A_0. Let*

$$f(Y^{-1})_0 = \sum_{\mu' \in L'} f_{\mu'}(X)t(-\mu').$$

Suppose that $\nu' \in L'$ is antidominant (i.e., $<\nu', \alpha> \leq 0$ *for all* $\alpha \in R^+$), *and that $f_{\mu'}(\nu' - \rho'_k) \neq 0$. Then $\mu' + \nu'$ is antidominant.*

Proof Let

$$f(Y^{-1}) = \sum_{\substack{\mu' \in L' \\ v \in W_0}} g_{\mu',v}(X)t(-\mu')v^{-1}$$

so that

$$f_{\mu'} = \sum_{v \in W_0} g_{\mu',v}.$$

By (2.8.2), $r'_k(\nu') = \nu' - \rho'_k$ since ν' is antidominant. Hence $f_{\mu'}(r'_k(\nu')) \neq 0$ and therefore $g_{\mu',v}(r'_k(\nu')) \neq 0$ for some $v \in W_0$. By (4.5.3) we have

(1) $$vt(\mu')(r'_k(\nu')) = r'_k(vt(\mu')\nu').$$

Let $\pi' = v(\mu' + \nu')$. Then the left-hand side of (1) is equal to $vt(\mu')(\nu' - \rho'_k) = \pi' - v\rho'_k$, and the right-hand side is

$$r'_k(\pi') = u(\pi')(-\rho'_k) = t(\pi')v(\pi')^{-1}(-\rho'_k) = \pi' - v(\pi')^{-1}\rho'_k.$$

Hence $v\rho'_k = v(\pi')^{-1}\rho'_k$ and so by (1.5.5) $v = v(\pi')^{-1}$. Consequently $\mu' + \nu' = v^{-1}\pi' = v(\pi')\pi'$ is antidominant. □

(4.5.8) *Let* $f \in A[\mathbf{c}]$. *If* $f(\lambda' + \rho'_k) = 0$ *for all regular dominant* $\lambda' \in L'$ (i.e., *such that* $<\lambda', \alpha> > 0$ *for all* $\alpha \in R^+$), *then* $f = 0$.

Proof By clearing denominators we may assume that $f \in A$, say

$$f = \sum_{i=1}^{r} f_i e^{\mu_i}$$

where $f_i \in K$ and $\mu_i \in L$. Let $\lambda' \in L'$ be dominant regular and such that the r numbers $<\lambda', \mu_i>$ are all distinct (we have only to avoid a finite number of hyperplanes $<\lambda', \mu_i - \mu_j> = 0$). Then

$$\sum_{i=1}^{r} f_i q^{<m\lambda' + \rho'_k, \mu_i>} = f(m\lambda' + \rho'_k) = 0$$

for all integers $m \geq 1$, and hence the polynomial

$$F(x) = \sum_{i=1}^{r} f_i q^{<\rho'_k, \mu_i>} x^{<\lambda', \mu_i>}$$

vanishes for infinitely many values of x, namely $x = q^m$, $m \geq 1$. Hence $F(x)$ is identically zero and so $f_1 = \cdots = f_r = 0$, i.e., $f = 0$. □

4.6 The operators $Y^{\lambda'}$

For each $a \in S_1$, let

(4.6.1) $$G_a = \tau_a + \mathbf{b}_a(X^{-1})(s_a - 1) = \mathbf{c}_a(X^{-1}) + \mathbf{b}_a(X^{-1})s_a$$

so that in particular $G_{a_i} = s_i T_i$ by (4.3.14). Clearly we have

(4.6.2) $$wG_a w^{-1} = G_{wa}$$

for all $w \in W$, and

(4.6.3) $$G_a^{-1} = \mathbf{c}_a(X) - \mathbf{b}_a(X^{-1})s_a$$

by use of (4.2.3) (iv).

Let $w \in W$ and let $w = u_j s_{i_1} \cdots s_{i_p}$ be a reduced expression. As in (2.2.9) let

$$b_r = s_{i_p} \cdots s_{i_{r+1}}(a_{i_r})$$

for $1 \leq r \leq p$, so that $S_1(w) = \{b_1, \ldots, b_p\}$. Then we have

(4.6.4) $$T(w) = wG_{b_1} \cdots G_{b_p}.$$

Proof From (4.6.2) it follows that

$$G_{b_r} = s_{i_p} \cdots s_{i_r} T_{i_r} s_{i_{r+1}} \cdots s_{i_p}$$

and therefore

$$T(w) = u_j T_{i_1} \cdots T_{i_p} = w G_{b_1} \cdots G_{b_p}. \qquad \square$$

Recall (2.8.3) that for $x \in \mathbb{R}$, $\eta(x) = 1$ if $x > 0$ and $\eta(x) = -1$ if $x \leq 0$.

(4.6.5) *Let $a \in S_1$ be such that $\alpha = Da$ is positive. Then for all $\mu \in L$ we have*

$$G_a e^{\mu} = \tau_a^{-\eta(<\mu,\alpha^\vee>)} e^{\mu} + \text{lower terms}.$$

Proof We have

(1) $$G_a e^{\mu} = \tau_a e^{\mu} + \boldsymbol{b}(\tau_a, \tau_a'; e^{-a})(e^{s_a \cdot \mu} - e^{\mu}).$$

If $<\mu, \alpha^\vee> = r > 0$, then $s_a \cdot \mu = \mu - ra$, and the right-hand side of (1) is equal to

$$\tau_a e^{\mu} - \sum_{j=0}^{r-1} u_j e^{\mu - ja}$$

where

$$u_j = \begin{cases} \tau_a - \tau_a^{-1} & \text{if } j \text{ is even,} \\ \tau_a' - \tau_a'^{-1} & \text{if } j \text{ is odd.} \end{cases}$$

Since $\mu - j\alpha < \mu$ for $1 \leq j \leq r - 1$, it follows that the leading term of $G_a e^{\mu}$ is $\tau_a^{-1} e^{\mu}$, which establishes (4.6.5) in this case.

If on the other hand $<\mu, \alpha^\vee> = -r < 0$, the right-hand side of (1) is now

$$\tau_a e^{\mu} + \sum_{j=1}^{r} u_j e^{\mu + ja}.$$

We have $\mu + j\alpha \in \Sigma^0(\mu)$ for $1 \leq j \leq r-1$ and $\mu + r\alpha = s_\alpha \mu < \mu$ by (2.7.9), since α is positive. Hence $\mu + j\alpha < \mu$ for $1 \leq j \leq r$, and the leading term of $G_a e^{\mu}$ is now $\tau_a e^{\mu}$.

Finally, it $<\mu, \alpha^\vee> = 0$ then $s_a e^{\mu} = e^{\mu}$, and hence $G_a e^{\mu} = \tau_a e^{\mu}$ by (4.6.1). $\square$

(4.6.6) *Suppose that $w \in W$ is such that Da is positive for all $a \in S(w)$. Then*

$$w^{-1} T(w) e^{\mu} = \tau(w, \mu) e^{\mu} + \text{lower terms},$$

where

(4.6.7) $$\tau(w,\mu) = \prod_{a\in S_1(w)} \tau_a^{-\eta(<\mu, Da^\vee>)}$$

Proof This follows from (4.6.4) and (4.6.5). □

For each $a \in S_1$ define

$$\kappa_a = \kappa_i$$

if $a \in Wa_i$, so that $\tau_a = q^{k_a/2}$. Then

(4.6.8) $$\tau(w,\mu) = q^{f(w,\mu)}$$

where

(4.6.9) $$f(w,\mu) = -\frac{1}{2}\sum_{a\in S_1(w)} \eta(<\mu, Da^\vee>)\kappa_a.$$

(4.6.10) *Let* $\lambda' \in L'_{++}$, $\mu \in L$. *Then*

$$f(t(\lambda'),\mu) = <\lambda', \mu - r_{k'}(\mu)>.$$

Proof Suppose first that $S = S(R)$ (1.4.1). Then for $\alpha \in R$

$$t(\lambda')(\alpha + rc) = \alpha + (r - <\lambda', \alpha>)c$$

so that

$$S(t(\lambda')) = \{\alpha + rc : \alpha \in R^+, 0 \le r < <\lambda', \alpha>\}$$

and therefore

$$\begin{aligned} f(t(\lambda'),\mu) &= -\frac{1}{2}\sum_{\alpha\in R^+} \eta(<\mu, \alpha^\vee>)<\lambda', \alpha>k(\alpha) \\ &= <\lambda', \mu - r_{k'}(\mu)> \end{aligned}$$

by (2.8.2), since $\alpha^\vee = \alpha'$ and $k(\alpha) = k'(\alpha^\vee)$.

Next, suppose that $S = S(R)^\vee$ (1.4.2). Then

$$S(t(\lambda')) = \{(\alpha + rc)^\vee : \alpha \in R^+, 0 \le r < <\lambda', \alpha>\}$$

and hence

$$\begin{aligned} f(t(\lambda'),\mu) &= -\frac{1}{2}\sum_{\alpha\in R^+} \eta(<\mu, \alpha>)\ <\lambda', \alpha>k(\alpha^\vee) \\ &= <\lambda', \mu - r_{k'}(\mu)> \end{aligned}$$

since now $\alpha' = \alpha$ and $k'(\alpha^\vee) = k(\alpha^\vee)$.

Finally, suppose that S is of type $(C_n^\vee, C_n)$ (1.4.3), so that $S_1 = S(R)^\vee$ where R is of type C_n. If $\alpha \in R$ is a long root, then $(\alpha + rc)^\vee$ is in the W-orbit of $a_n = \alpha_n^\vee$ if r is even, and in the W-orbit of a_0 if r is odd; moreover, in this case $<\lambda', \alpha>$ is an even integer. It follows that if $\alpha \in R^+$ is a long root, the contribution to $f(t(\lambda'), \mu)$ from the roots $(\alpha + rc)^\vee$ in $S_1(t(\lambda'))$ is

$$-\frac{1}{4}\eta(<\mu, \alpha>) <\lambda', \alpha>(\kappa_n + \kappa_0) = -\frac{1}{2}\eta(<\mu, \alpha'>) <\lambda', \alpha>k'(\alpha^\vee),$$

since $\kappa_n + \kappa_0 = k_1 + k_2 + k_3 + k_4 = 2k_1' = 2k'(\alpha^\vee)$. Hence again we have

$$f(t(\lambda'), \mu) = <\lambda', \mu - r_{k'}(\mu)>. \qquad \square$$

(4.6.11) *Let $\lambda' \in L'$, $\mu \in L$. Then*

$$Y^{\lambda'} e^\mu = q^{-<\lambda', r_{k'}(\mu)>} e^\mu + \text{lower terms}.$$

Proof Suppose first that λ' is dominant. Then $Y^{\lambda'} = T(t(\lambda'))$ and $t(\lambda')$ satisfies the conditions of (4.6.6), and therefore

(1) $$t(\lambda')^{-1} Y^{\lambda'} e^\mu = q^{<\lambda', \mu - r_{k'}(\mu)>} e^\mu + \text{ lower terms}$$

by virtue of (4.6.6), (4.6.8) and (4.6.10). Since

$$t(\lambda') e^\mu = q^{-<\lambda', \mu>} e^\mu,$$

it follows from (1) that (4.6.11) is true when λ' is dominant.

If now λ' is not dominant, then $\lambda' = \lambda_1' - \lambda_2'$ with λ_1', λ_2' both dominant. Hence

$$\begin{aligned} Y^{\lambda'} e^\mu &= Y^{\lambda_1'} (Y^{\lambda_2'})^{-1} e^\mu \\ &= q^{<\lambda_2' - \lambda_1', r_{k'}(\mu)>} e^\mu + \text{ lower terms} \\ &= q^{-<\lambda', r_{k'}(\mu)>} e^\mu + \text{ lower terms}. \end{aligned} \qquad \square$$

If we regard each $f \in A'$ as a function on V as in (4.5.1), we may restate (4.6.11) as follows:

(4.6.12) *Let $f \in A'$ and $\mu \in L$. Then*

$$f(Y) e^\mu = f(-r_{k'}(\mu)) e^\mu + \text{lower terms}.$$

(4.6.13) *Let $f \in A_0'$ and $\mu \in L_{++}$. Then*

$$f(Y) m_\mu = f(-\mu - \rho_{k'}) m_\mu + \text{lower terms}.$$

For $-r_{k'}(w_0\mu) = -w_0\mu + \rho_{k'} = -w_0(\mu + \rho_{k'})$. $\square$

4.7 The double affine Hecke algebra

Suppose first that S is reduced ((1.4.1), (1.4.2)). Then the *double affine Hecke algebra* $\tilde{\mathfrak{H}}$ is the quotient of the group algebra $K\tilde{\mathfrak{B}}$ of the double braid group $\tilde{\mathfrak{B}}$ by the ideal generated by the elements

$$(T_i - \tau_i)(T_i + \tau_i^{-1}) \qquad (i \in I).$$

Thus $\tilde{\mathfrak{H}}$ is generated over K by $\mathfrak{H}$ and $X^L = \{X^\lambda : \lambda \in L\}$, subject to the relations (3.4.2)–(3.4.5).

Suppose now that S is of type $(C_n^\vee, C_n)$ (1.4.3). Let

(4.7.1) $$T_n' = X^{-a_n} T_n^{-1}, \quad T_0' = X^{-a_0} T_0^{-1}$$

as in (4.3.19). Then in this case the double affine Hecke algebra $\tilde{\mathfrak{H}}$ is the quotient of $K\tilde{\mathfrak{B}}$ by the ideal generated by the elements

$$(T_i - \tau_i)(T_i + \tau_i^{-1}) \ (0 \le i \le n), (T_0' - \tau_0')(T_0' + \tau_0'^{-1}), \ (T_n' - \tau_n')(T_n' + \tau_n'^{-1}).$$

Thus $\tilde{\mathfrak{H}}$ is generated over K by $\mathfrak{H}$ and X^L subject to the relations (3.4.2)–(3.4.5) and

(4.7.2) $$(T_0' - \tau_0')(T_0' + \tau_0'^{-1}) = (T_n' - \tau_n')(T_n' + \tau_n'^{-1}) = 0.$$

Let $f \in \Lambda = L \oplus \mathbb{Z}c_0$ (1.4.8). Then in $\tilde{\mathfrak{H}}$ we have

(4.7.3) $$T_i X^f - X^{s_i f} T_i = \boldsymbol{b}_i(X)(X^f - X^{s_i f})$$

for all $i \in I$,

Proof If $<L, a_i^\vee> = \mathbb{Z}$, this follows from the relations (3.4.2)–(3.4.4) in the same way that (4.2.4) was a consequence of (3.2.4).

If $<L, a_i^\vee> = 2\mathbb{Z}$, so that we are in the situation of (1.4.3) and $i = 0$ or n, then just as in the proof of (4.2.4) it is enough to verify (4.7.3) for a single $f \in \Lambda$ such that $<f, a_i^\vee> = 2$. We take $f = a_i$ ($i = 0$ or n) and calculate

$$\begin{aligned} T_i X^{a_i} - X^{-a_i} T_i &= T_i X^{a_i} - X^{-a_i}(T_i^{-1} + \tau_i - \tau_i^{-1}) \\ &= T_i'^{-1} - T_i' - (\tau_i - \tau_i^{-1})X^{-a_i} \\ &= -(\tau_i' - \tau_i'^{-1}) - (\tau_i - \tau_i^{-1})X^{-a_i} \\ &= \boldsymbol{b}_i(X)(X^{a_i} - X^{-a_i}) \end{aligned}$$

by use of (4.7.1) and (4.7.2). □

From (4.7.3) and (4.3.15) it follows that the representation β of $\mathfrak{H}$ on A (4.3.10) extends to a representation (also denoted by β) of $\tilde{\mathfrak{H}}$ on A, such that $\beta(X^\mu)$ is multiplication by e^μ for $\mu \in L$.

(4.7.4) (i) *The representation β of $\tilde{\mathfrak{H}}$ on A is faithful*
(ii) *The elements $T(w)X^\mu$ (resp. the elements $X^\mu T(w)$), where $w \in W$ and $\mu \in L$, form a K-basis of $\tilde{\mathfrak{H}}$.*

Proof As in the proof of (4.2.7), it follows from (4.7.3) that the elements $T(w)X^\mu$ (resp. $X^\mu T(w)$) span $\tilde{\mathfrak{H}}$ as a K-vector space. On the other hand, by (4.3.11), their images under β are linearly independent as linear operators on A. This proves both parts of (4.7.4). □

(4.7.5) *The elements $Y^{\lambda'}T(w)X^\mu$ (resp. the elements $X^\mu T(w)Y^{\lambda'}$) where $\lambda' \in L'$, $\mu \in L$ and $w \in W_0$, from a K-basis of $\tilde{\mathfrak{H}}$.*

This follows from (4.2.7) and (4.7.4). □

Now let $\tilde{\mathfrak{H}}'$ be the algebra defined as follows. If S is reduced ((1.4.1), (1.4.2)), $\tilde{\mathfrak{H}}'$ is obtained from $\tilde{\mathfrak{H}}$ by interchanging R and R', L and L'. If S is of type $(C_n^\vee, C_n)$ (1.4.3), $\tilde{\mathfrak{H}}'$ is obtained from $\tilde{\mathfrak{H}}$ by interchanging the parameters τ_0 and τ_n' (which affects only the relations (4.7.2)).

(4.7.6) *The K-linear mapping ω: $\tilde{\mathfrak{H}}' \to \tilde{\mathfrak{H}}$ defined by*

$$\omega(X^{\lambda'}T(w)Y^\mu) = X^{-\mu}T(w^{-1})Y^{-\lambda'}$$

($\lambda' \in L'$, $\mu \in L$, $w \in W_0$) is an anti-isomorphism of K-algebras.

Proof In view of the duality theorem (3.5.1) we have only to verify that ω respects the Hecke relations in $\tilde{\mathfrak{H}}'$ and $\tilde{\mathfrak{H}}$.
(a) In the case where S is reduced ((1.4.1), (1.4.2)) we have to show that T_0^* defined by (3.5.2) satisfies

$$(T_0^* - \tau_0)(T_0^* + \tau_0^{-1}) = 0.$$

From (3.4.9) it follows that T_0^* is conjugate in $\tilde{\mathfrak{B}}$ to $T_i^{-1}X^{-a_i}$ for some $i \neq 0$ such that $\tau_i = \tau_0$. Hence it is enough to show that

$$T_i^{-1}X^{-a_i} - X^{a_i}T_i = \tau_i - \tau_i^{-1}$$

or equivalently that

$$T_i X^{-a_i} - X^{a_i} T_i = \left(\tau_i - \tau_i^{-1}\right)(1 + X^{-a_i}).$$

But this is the case $f = -a_i$ of (4.7.3).

(b) When S is of type $(C_n^\vee, C_n)$ (1.4.3) we have to show that

(1) $$(\omega(T_0) - \tau_n')\left(\omega(T_0) + \tau_n'^{-1}\right) = 0,$$

(2) $$(\omega(T_0') - \tau_0')\left(\omega(T_0') + \tau_0'^{-1}\right) = 0,$$

(3) $$(\omega(T_n') - \tau_0)\left(\omega(T_n') + \tau_0^{-1}\right) = 0,$$

By (3.4.9), $\omega(T_0) = T_0^*$ is conjugate in $\tilde{\mathfrak{B}}$ to $T_n^{-1} X^{-a_n}$, hence also to $X^{-a_n} T_n^{-1} = T_n'$, which proves (1). Next, we have

$$\begin{aligned}\omega(T_0') &= \omega\left(q^{-1/2} X^{\varepsilon_1} T_0^{-1}\right) = q^{-1/2} \omega(T_0)^{-1} Y^{-\varepsilon_1} \\ &= q^{-1/2} (Y^{-\varepsilon_1} T_0 X^{-\varepsilon_1})^{-1} Y^{-\varepsilon_1} = q^{-1/2} X^{\varepsilon_1} T_0^{-1} = T_0',\end{aligned}$$

which proves (2). Finally, by (3.3.7), $\omega(T_n') = T_n^{-1} Y^{\varepsilon_n}$ is conjugate to T_0, which gives (3). □

By (4.7.4) we may identify $\tilde{\mathfrak{H}}$ with its image under β, and regard each $h \in \tilde{\mathfrak{H}}$ as a linear operator on A. We define a K-linear map $\theta: \tilde{\mathfrak{H}} \to K$ as follows:

(4.7.7) $$\theta(h) = h(1_A)(-\rho_k'),$$

where 1_A is the identity element of A. Dually, we define $\theta' : \tilde{\mathfrak{H}}' \to K$ by

(4.7.7′) $$\theta'(h') = h'(1_{A'})(-\rho_{k'}).$$

Suppose that $h = f(X)T(w)g(Y^{-1})$, where $f \in A$, $g \in A'$ and $w \in W_0$. By (4.6.12) we have

(1) $$g(Y^{-1})(1_A) = g(-\rho_{k'})1_A.$$

If $w = s_{i_1} \cdots s_{i_p}$, let $\tau_w = \tau_{i_1} \cdots \tau_{i_p}$. Since $T_i(1_A) = \tau_i 1_A$, it follows that

(2) $$T(w)1_A = \tau_w 1_A.$$

From (1) and (2) we have

(4.7.8) $$\theta(f(x)T(w)g(Y^{-1})) = f(-\rho_k')\tau_w g(-\rho_{k'}).$$

Since

$$f(X)T(w)g(Y^{-1}) = \omega(g(X)T(w^{-1})f(Y^{-1}))$$

it follows from (4.7.5) and (4.7.8) that

$$\theta' = \theta \circ \omega. \tag{4.7.9}$$

Next, if $h \in \tilde{\mathfrak{H}}$ and $h' \in \tilde{\mathfrak{H}}'$, we define

$$[h, h'] = \theta'(\omega^{-1}(h)h'), \tag{4.7.10}$$

$$[h', h] = \theta(\omega(h')h). \tag{4.7.10$'$}$$

From (4.7.9) it follows that

$$[h', h] = [h, h']. \tag{4.7.11}$$

Also, if $h_1 \in \tilde{\mathfrak{H}}$ we have

$$[h_1 h, h'] = [h, \omega^{-1}(h_1)h'] \tag{4.7.12}$$

because $\theta'(\omega^{-1}(h_1 h)h') = \theta'(\omega^{-1}(h)\omega^{-1}(h_1)h')$.

In particular, if $f \in A$ and $f' \in A'$ we define

$$[f, f'] = [f(X), f'(X)] = \theta'(f(Y^{-1})f'(X)) = (f(Y^{-1})f')(-\rho_{k'}) \tag{4.7.13}$$

and dually

$$[f', f] = [f'(X), f(X)] = (f'(Y^{-1})f)(-\rho'_k). \tag{4.7.13$'$}$$

From (4.7.11), this pairing between A and A' is symmetric:

$$[f', f] = [f, f']. \tag{4.7.14}$$

Notes and references

The fundamental relation (4.2.4) is Lusztig's Prop. 3.6 of [L1]. The basic representation β of the affine Hecke algebra $\mathfrak{H}$ and its properties are due to Cherednik [C2], as is the double affine Hecke algebra $\tilde{\mathfrak{H}}$. The mappings θ, θ' and the pairing (4.7.10) are also due to Cherednik [C4].

5

Orthogonal polynomials

5.1 The scalar product

Let S be an irreducible affine root system, as in Chapter 1. Fix a basis $(a_i)_{i\in I}$ of S, and let S^+ be the set of positive affine roots determined by this basis. Let

$$S_1 = \left\{a \in S : \tfrac{1}{2}a \notin S\right\}$$

as in §1.3, so that $S_1 = S$ if S is reduced, and in any case S_1 is a reduced affine root system with the same basis (a_i) as S.

As in §1.2 there is a unique relation of the form

$$\sum_{i\in I} m_i a_i = c$$

where the m_i are positive integers with no common factor, and c is a positive constant function. Fix an index $0 \in I$ as in §1.2 (so that $m_0 = 1$).

We shall in fact assume that S is as in (1.4.1), (1.4.2) or (1.4.3). This assumption excludes the reduced affine root systems of type BC_n (1.3.6) and the non-reduced systems other than $(C_n^\vee, C_n)$. The reason for this exclusion will become apparent later (5.1.7).

As in §1.4, let

$$\Lambda = L \oplus \mathbb{Z}c_0$$

and let

$$\Lambda^+ = L \oplus \mathbb{N}c_0.$$

The affine roots $a \in S$ lie in Λ, and the positive affine roots in Λ^+. If $f \in \Lambda$, say $f = \mu + rc_0$, where $\mu \in L$ and $r \in \mathbb{Z}$, let

$$e^f = q_0^r e^\mu = q^{r/e} e^\mu$$

as in §4.3 (where $q_0 = q^{1/e}$).

For each $a \in S$ let t_a be a positive real number such that $t_a = t_b$ if a, b are in the same W-orbit in S, where W is the extended affine Weyl group of S. The t_a determine a labelling k of S as follows: if $a \in S_1$,

(5.1.1) $$q^{k(a)} = t_a t_{2a}^{1/2}, \quad q^{k(2a)} = t_{2a}^{1/2}$$

where $t_{2a}^{1/2}$ is the positive square root, and $t_{2a} = 1$ (so that $k(2a) = 0$) if $2a \notin S$.

For each $a \in S$ let

(5.1.2) $$\Delta_a = \Delta_{a,k} = \frac{1 - t_{2a}^{1/2} e^a}{1 - t_a t_{2a}^{1/2} e^a}.$$

If $a \in S_1$ we have

$$\begin{aligned} \Delta_a \Delta_{2a} &= \frac{(1 - t_{2a}^{1/2} e^a)(1 - e^{2a})}{(1 - t_a t_{2a}^{1/2} e^a)(1 - t_{2a} e^{2a})} \\ &= \frac{1 - e^{2a}}{\left(1 - q^{k(a)} e^a\right)\left(1 + q^{k(2a)} e^a\right)} \end{aligned}$$

so that

(5.1.3) $$(\Delta_a \Delta_{2a})^{-1} = \tau_a \mathbf{c}(\tau_a, \tau'_a; e^a)$$

in the notation of (4.2.2), where

(5.1.4) $$\tau_a = (t_a t_{2a})^{1/2} = q^{\kappa_a/2}, \quad \tau'_a = t_a^{1/2} = q^{\kappa'_a/2}$$

(so that $\tau_a = \tau'_a$ if $2a \notin S$), and

(5.1.5) $$\kappa_a = k(a) + k(2a), \quad \kappa'_a = k(a) - k(2a)$$

as in §4.4. Let

(5.1.6) $$\tau_i = q^{\kappa_i/2} = \tau_{a_i}, \quad \tau'_i = q^{\kappa'_i/2} = \tau'_{a_i}.$$

for each $i \in I$, so that $\kappa_i = \kappa'_i$ for all $i \in I$ except when S is of type $(C_n^\vee, C_n)$ and $i = 0$ or n.

We now define the weight function

(5.1.7) $$\Delta = \Delta_{S,k} = \prod_{a\in S^+} \Delta_a = \prod_{a\in S^+} \frac{1 - t_{2a}^{1/2} e^a}{1 - t_a t_{2a}^{1/2} e^a}.$$

We may remark that if S^0 is a subsystem of S, then $\Delta_{S^0,k}$ is obtained from $\Delta_{S,k}$ by setting $t_a = 1$ for all $a \in S - S^0$. Thus if S is of type $(C_n^\vee, C_n)$ and S^0 is one of the non-reduced systems (1.3.15)–(1.3.17), or one of the "classical" reduced systems (1.3.2)–(1.3.7), $\Delta_{S^0,k}$ is obtained from $\Delta_{S,k}$ by setting some of the t_a equal to 1.

On expansion, Δ is a formal power series in the exponentials e^{a_i} $(i \in I)$, with coefficients in the ring of polynomials in the t_a and $t_{2a}^{1/2}$: say

$$\Delta = \sum_{b\in\Lambda^+} u_b e^b = \sum_{\substack{\lambda\in L\\ r\geq 0}} u_{\lambda+rc} q^r e^\lambda.$$

If $f \in A$, say

$$f = \sum_{\lambda\in L} f_\lambda e^\lambda,$$

the *constant term* of $f\Delta$ is defined to be

(5.1.8) $$\mathrm{ct}(f\Delta) = \sum_{r\geq 0} \left(\sum_{\lambda\in L} u_{\lambda+rc} f_{-\lambda} \right) q^r,$$

a formal power series in q.

Let

(5.1.9) $$\Delta_1 = \Delta/\mathrm{ct}(\Delta) = \sum_{\mu\in L} v_\mu(q,t) e^\mu$$

so that $v_0(q,t) = 1$.

(5.1.10) (i) *The coefficients* $v_\mu(q,t)$ *are rational functions of* q *and the* t_a, $t_{2a}^{1/2}$.
(ii) $v_\mu(q,t) = v_{-\mu}(q^{-1}, t^{-1})$ *for all* $\mu \in L$.

Proof We shall give the proof when S is reduced. As in §1.4 we shall assume that $|\varphi|^2 = 2$, where φ is the highest root of R. For each $i \in I$, since s_i permutes $S^+ - \{a_i\}$, we have

(5.1.11) $$\frac{s_i \Delta_1}{\Delta_1} = \frac{1 - e^{-a_i}}{1 - t_i e^{-a_i}} \frac{1 - t_i e^{a_i}}{1 - e^{a_i}} = \frac{1 - t_i e^{a_i}}{t_i - e^{a_i}}$$

where $t_i = t_{a_i}$. Hence

(1) $$(1 - t_i e^{a_i}) \left(\sum v_\mu e^\mu \right) = (t_i - e^{a_i}) \sum v_\mu e^{s_i \mu}.$$

Suppose first that $i \neq 0$. Then by comparing coefficients of $e^{\mu + a_i}$ on either side of (1) we obtain

(2) $$v_{\mu + a_i} - t_i v_\mu = t_i v_{s_i \mu - a_i} - v_{s_i \mu}.$$

Suppose next that $i = 0$. Since $a_0 = -\varphi + c$ (because $|\varphi|^2 = 2$) we have $s_0 \cdot \mu = s_\varphi \mu + <\mu, \varphi> c$ and therefore

$$\sum_\mu v_\mu e^{s_0 \cdot \mu} = \sum_\mu q^{-<\mu,\varphi>} v_{s_\varphi \mu} e^\mu.$$

Hence by equating coefficients of $e^{\mu - \varphi}$ on either side of (1) we obtain

(3) $$v_{\mu - \varphi} - q t_0 v_\mu = q^{-<\mu,\varphi>+2} t_0 v_{s_\varphi \mu + \varphi} - q^{-<\mu,\varphi>+1} v_{s_\varphi \mu}.$$

Let Φ denote the field of rational functions in q and the t_i. We proceed by induction on μ, and assume that $v_\nu \in \Phi$ for all ν in a lower W_0-orbit than μ. Let λ be the dominant element of the orbit $W_0 \mu$, and suppose first that $\mu \neq \lambda$. Then for some $i \neq 0$ we have $<\mu, a_i^\vee> = -r < 0$, so that $s_i \mu = \mu + r a_i < \mu$ by (2.7.9). From (2) we obtain

$$v_\mu - t_i^{-1} v_{s_i \mu} \in \Phi$$

and hence by iteration, for all $\mu \in W_0 \lambda$,

(4) $$v_\mu - t_w^{-1} v_\lambda \in \Phi,$$

where $w \in W_0$ is the shortest element such that $w\mu = \lambda$, and $t_w = t_{i_1} \cdots t_{i_p}$ if $w = s_{i_1} \cdots s_{i_p}$ is a reduced expression.

Next, we have $<\lambda, \varphi> = r' \geq 1$ since λ is dominant. Hence from (3) applied to λ we obtain

(5) $$v_\lambda - t_0^{-1} q^{-r'} v_{s_\varphi \lambda} \in \Phi.$$

From (4) (with $\mu = s_\varphi \lambda$) and (5) it follows that $v_\lambda \in \Phi$ and hence that $v_\mu \in \Phi$ for all $\mu \in W_0 \lambda$.

This proves (5.1.10) (i) and shows that Δ_1 is uniquely determined by the relations (5.1.11), together with the fact that the constant term of Δ_1 is 1. Now these relations are unaltered by replacing t_i and e^{a_i} by t_i^{-1} and e^{-a_i}. Hence the same is true of Δ_1, which establishes (5.1.10) (ii).

Finally, when S is not reduced, the argument is essentially the same: (5.1.11) is replaced by

$$\frac{s_i \Delta_1}{\Delta_1} = \frac{(1 - t_i e^{a_i})(1 + t_i' e^{a_i})}{(t_i - e^{a_i})(t_i' + e^{a_i})}$$

where $t_i = t_{a_i} t_{2a_i}^{1/2}$ and $t_i' = t_{2a_i}^{1/2}$, and the recurrence relations (2) and (3) are correspondingly more complicated. □

(5.1.12) When $S = S(R)$ with R reduced, we have

$$\Delta = \prod_{\alpha \in R^+} \frac{(e^\alpha; q)_\infty (q e^{-\alpha}; q)_\infty}{\left(q^{k(\alpha)} e^\alpha; q\right)_\infty \left(q^{k(\alpha)+1} e^{-\alpha}; q\right)_\infty}$$

where we have made use of the standard notation

$$(x; q)_\infty = \prod_{i=0}^{\infty} (1 - x q^i).$$

Since $(x; q)_\infty / (q^k x; q)_\infty \to (1 - x)^k$ as $q \to 1$, for all $k \in \mathbb{R}$ (see e.g. [G1], Chapter 1), it follows that

$$\Delta \to \prod_{\alpha \in R} (1 - e^\alpha)^{k(\alpha)}$$

as $q \to 1$.

Next, if the labels $k(\alpha)$ are non-negative integers, Δ is a finite product, namely

$$\Delta = \prod_{\alpha \in R^+} (e^\alpha; q)_{k(\alpha)} (q e^{-\alpha}; q)_{k(\alpha)}$$

where

$$(x; q)_k = \prod_{i=0}^{k-1} (1 - x q^i)$$

for $k \in \mathbb{N}$. Equivalently,

$$\Delta = \prod_{a \in S(k)} (1 - e^a)$$

where

$$S(k) = \{a \in S : a(x) \in (0, k(a)) \text{ for } x \in C\}$$

and C is the fundamental alcove (§1.2) for W_S.

(5.1.13) Suppose next that $S = S(R)^\vee$, where R is reduced and $R^\vee \neq R$. As in §1.4, we assume that $|\alpha|^2 = 2$ if $\alpha \in R$ is a long root. Let $u_\alpha = 2/|\alpha|^2$ for

$\alpha \in R$; then $u_\alpha = 1$ if α is long and $u_\alpha = d$ if α is short, where d (= 2 or 3) is the maximum bond-strength in the Dynkin diagram of R. Let $q_\alpha = q^{u_\alpha}$ for each $\alpha \in R$, and let $k^\vee(\alpha) = u_\alpha^{-1} k(\alpha^\vee)$. Then we have

$$\Delta = \prod_{\alpha \in R^+} \frac{\left(e^{\alpha^\vee}; q_\alpha\right)_\infty \left(q_\alpha e^{-\alpha^\vee}; q_\alpha\right)_\infty}{\left(q_\alpha^{k^\vee(\alpha)} e^{\alpha^\vee}; q_\alpha\right)_\infty \left(q_\alpha^{k^\vee(\alpha)+1} e^{-\alpha^\vee}; q_\alpha\right)_\infty},$$

and $\Delta \to \prod_{\alpha \in R}(1 - e^{\alpha^\vee})^{k^\vee(\alpha)}$ as $q \to 1$.

If each $k^\vee(\alpha)$ is a non-negative integer,

$$\begin{aligned} \Delta &= \prod_{\alpha \in R^+} (e^{\alpha^\vee}; q_\alpha)_{k^\vee(\alpha)} (q_\alpha e^{-\alpha^\vee}; q_\alpha)_{k^\vee(\alpha)} \\ &= \prod_{a \in S(k)} (1 - e^a) \end{aligned}$$

with $S(k)$ as defined in the previous paragraph (5.1.12).

(5.1.14) When S is of type $(C_n^\vee, C_n)$ and $W = W_S$ is the affine Weyl group of type C_n, the orbits $O_1, \ldots, O_5$ of W in S were described in §1.3. In the notation of (1.3.18) let

$$\begin{aligned} R_1 &= \{\pm\varepsilon_1, \ldots, \pm\varepsilon_n\}, \quad & R_2 &= \{\pm\varepsilon_i \pm \varepsilon_j : 1 \le i < j \le n\}, \\ R_1^+ &= \{\varepsilon_1, \ldots, \varepsilon_n\}, & R_2^+ &= \{\varepsilon_i \pm \varepsilon_j : 1 \le i < j \le n\}, \end{aligned}$$

and let

$$\begin{aligned} (u_1, \ldots, u_4) &= \left(q^{k_1}, -q^{k_2}, q^{k_3+\frac{1}{2}}, -q^{k_4+\frac{1}{2}}\right), \\ (u_1', \ldots, u_4') &= \left(q^{k_1+1}, -q^{k_2+1}, q^{k_3+\frac{1}{2}}, -q^{k_4+\frac{1}{2}}\right) \end{aligned}$$

where $k_i = k(a)$ for $a \in O_i$, as in §1.5. Then $\Delta = \Delta^{(1)}\Delta^{(2)}$, where

$$\Delta^{(1)} = \prod_{\alpha \in R_1^+} \frac{(e^{2\alpha}; q)_\infty (q e^{-2\alpha}; q)_\infty}{\prod_{i=1}^4 (u_i e^\alpha; q)_\infty (u_i' e^{-\alpha}; q)_\infty},$$

and

$$\Delta^{(2)} = \prod_{\alpha \in R_2^+} \frac{(e^\alpha; q)_\infty (q e^{-\alpha}; q)_\infty}{(q^{k_5} e^\alpha; q)_\infty (q^{k_5+1} e^{-\alpha}; q)_\infty}.$$

When $q \to 1$ we have

$$\begin{aligned} \Delta^{(1)} &\to \prod_{\alpha \in R_1} (1 - e^\alpha)^{k_1+k_3} (1 + e^\alpha)^{k_2+k_4}, \\ \Delta^{(2)} &\to \prod_{\alpha \in R_2} (1 - e^\alpha)^{k_5}. \end{aligned}$$

If each of $k_1, \dots, k_4$ is a non-negative integer, then

$$\Delta^{(1)} = \prod_{\alpha \in R_1^+} \prod_{i=1}^{4} (v_i e^{\alpha}; q)_{k_i} (v_i' e^{-\alpha}; q)_{k_i}$$

where

$$(v_1, \dots, v_4) = \left(1, -1, q^{1/2}, -q^{1/2}\right),$$
$$(v_1', \dots, v_4') = \left(q, -q, q^{1/2}, -q^{1/2}\right).$$

Let K be the field generated over $\mathbb{Q}$ by the τ_a, τ_a' ($a \in S$) and $q_0 = q^{1/e}$, and as in Chapter 4 let $A = KL$ and $A' = KL'$ denote the group algebras of L and L' over K. We define an involution $f \mapsto f^*$ on A and on A' as follows: if

$$f = \sum_{\lambda} f_\lambda e^{\lambda}$$

with coefficients $f_\lambda \in K$, then

(5.1.15) $$f^* = \sum_{\lambda} f_\lambda^* e^{-\lambda}$$

where f_λ^* is obtained from f_λ by replacing q_0, τ_a, τ_a' by their inverses q_0^{-1}, τ_a^{-1}, $\tau_a'^{-1}$ respectively. Thus for example

$$(e^a)^* = e^{-a}$$

for all $a \in S$.

If the labels $k(\alpha)$ in (5.1.12), $k^\vee(\alpha)$ in (5.1.13) and k_i in (5.1.14) are non-negative integers, so that Δ is a finite product and hence an element of A, then

(5.1.16) $$\Delta^* = q^{-N(k)} \Delta,$$

where

$$N(k) = \begin{cases} \sum_{\alpha \in R^+} k(\alpha)^2 & \text{if } S = S(R), \\ \sum_{\alpha \in R^+} u_\alpha k^\vee(\alpha)^2 & \text{if } S = S(R)^\vee, \\ n\left(k_1^2 + \cdots + k_4^2\right) + n(n-1)k_5^2 & \text{if } S = \left(C_n^\vee, C_n\right). \end{cases}$$

Proof This is a matter of simple calculation. For example, if $S = S(R)$, then by (5.1.12)

$$\Delta = \prod_{\alpha \in R^+} \prod_{i=0}^{k(\alpha)-1} (1 - q^i e^{\alpha})(1 - q^{i+1} e^{-\alpha})$$

so that

$$\begin{aligned}\Delta^* &= \prod_{\alpha\in R^+}\prod_{i=0}^{k(\alpha)-1} q^{-2i-1}(1-q^i e^{\alpha})(1-q^{i+1}e^{-\alpha})\\ &= q^{-N(k)}\Delta,\end{aligned}$$

where

$$N(k) = \sum_{\alpha\in R^+}\sum_{i=0}^{k(\alpha)-1}(2i+1) = \sum_{\alpha\in R^+} k(\alpha)^2.$$

Similarly in the other cases. □

We now define a scalar product on A as follows:

(5.1.17) $$(f, g) = \mathrm{ct}(fg^*\Delta)$$

where $f, g \in A$ and $\Delta = \Delta_{S,k}$, and as before ct means constant term. This scalar product is sesquilinear, i.e.

$$(\xi f, g) = \xi(f, g), \quad (f, \xi g) = \xi^*(f, g)$$

for $\xi \in K$. We shall also define the normalized scalar product

(5.1.18) $$(f, g)_1 = \mathrm{ct}(fg^*\Delta_1) = (f, g)/(1, 1).$$

This normalized scalar product is K-valued and Hermitian, i.e.,

(5.1.19) $$(g, f)_1 = (f, g)_1^*$$

for $f, g \in A$. For if

$$f = \sum f_\lambda e^\lambda, \quad g = \sum g_\mu e^\mu, \quad \Delta_1 = \sum v_\lambda e^\lambda$$

then $v_\lambda = v_{-\lambda}^* \in K$ by (5.1.10), and

$$\begin{aligned}(g, f)_1 &= \sum_{\lambda,\mu} f_\lambda^* g_\mu v_{\lambda-\mu} = \left(\sum_{\lambda,\mu} f_\lambda g_\mu^* v_{\mu-\lambda}\right)^*\\ &= (f, g)_1^*.\end{aligned}$$

□

Dually, we define a scalar product on A' by

(5.1.17′) $$(f, g)' = \mathrm{ct}(fg^*\Delta')$$

where $f, g \in A'$ and $\Delta' = \Delta_{S',k'}$, and S', k' are as defined in §1.4 and §1.5.

(5.1.20) *Let* $f \in A$, $f \neq 0$. *Then* (f, f) *is not identically zero.*

Proof Let

$$f = \sum_{\mu} f_{\mu}(q, t)e^{\mu}.$$

As functions of q_0, each coefficient f_μ has only finitely many zeros and poles. Hence we can choose labels $k(a) \in \mathbb{N}$ for which f is well-defined and nonzero. By multiplying f by a suitable power of $1 - q_0$, we may further assume that f is well-defined and nonzero at $q_0 = 1$. The coefficients f_μ are now rational numbers. Now when $q_0 = 1$ it follows from (5.1.12)–(5.1.14) that $\Delta = FF^*$, where F is the product of a finite number of factors of the form $1 \pm e^{\alpha}$, $\alpha \in R$. Let

$$g = Ff = \sum g_{\lambda} e^{\lambda}$$

with coefficients $g_\lambda \in \mathbb{Q}$. Then

$$(f, f) = \text{ct}(gg^*) = \sum g_{\lambda}^2 > 0$$

and so (f, f) is not identically zero as a function of q_0 and the t's. □

From (5.1.20) it follows that

(5.1.21) *The restriction of the scalar product* (f, g) *to every nonzero subspace of* A *is nondegenerate.*

If $F : A \to A$ is a linear operator, we denote by F^* the adjoint of F (when it exists), so that

$$(Ff, g) = (f, F^*g)$$

for all $f, g \in A$.

Let $\tilde{\mathfrak{H}}$ be the double affine Hecke algebra (§4.7), identified via the representation β (4.7.4) with a ring of operators on A. We recall that the elements $T(w)f(X)$ $(w \in W, f \in A)$ form a K-basis of $\tilde{\mathfrak{H}}$.

(5.1.22) *Each* $F \in \tilde{\mathfrak{H}}$ *has an adjoint* F^*. *If* $F = T(w)f(X)$ *then* $F^* = f^*(X)T(w)^{-1}$.

Proof It is clear from the definitions that the adjoint of $f(X)$ is $f^*(X)$, and that the adjoint of U_j $(= u_j)$ is u_j^{-1} $(j \in J)$. Hence it is enough to show that

$T_i^* = T_i^{-1}$ for each $i \in I$. If $f, g \in A$ we have

$$\begin{aligned}(s_i f, g) &= \mathrm{ct}((s_i f) g^* \Delta) \\ &= \mathrm{ct}(f (s_i g)^* s_i \Delta)\end{aligned}$$

and by (5.1.3)

$$s_i \Delta = \frac{\boldsymbol{c}_i(X)}{\boldsymbol{c}_i(X^{-1})} \Delta$$

where $\boldsymbol{c}_i(X^{\pm 1}) = \boldsymbol{c}_i(\tau_i, \tau_i'; X^{\pm a_i})$. It follows that the adjoint of s_i is

(5.1.23)
$$s_i^* = \frac{\boldsymbol{c}_i(X)}{\boldsymbol{c}_i(X^{-1})} s_i .$$

Since (4.3.12)

$$T_i = \tau_i + \boldsymbol{c}_i(X)(s_i - 1)$$

it follows that T_i has an adjoint and that (since $\boldsymbol{c}_i^* = \boldsymbol{c}_i$)

$$\begin{aligned}T_i^* &= \tau_i^{-1} + (s_i^* - 1)\boldsymbol{c}_i(X) \\ &= \tau_i^{-1} + \left(\frac{\boldsymbol{c}_i(X)}{\boldsymbol{c}_i(X^{-1})} s_i - 1 \right) \boldsymbol{c}_i(X) \\ &= \tau_i^{-1} + \boldsymbol{c}_i(X)(s_i - 1) = T_i^{-1}.\end{aligned}$$

□

In particular:

(5.1.24) *If $f \in A'$, the adjoint of $f(Y)$ is $f^*(Y)$.* □

Later we shall require a symmetric variant of the scalar product (5.1.17). Let

$$S_0 = \{a \in S : a(0) = 0\},$$

which is a finite root system, let $S_0^+ = S_0 \cap S^+$, and define

(5.1.25)
$$\Delta^0 = \Delta_{S,k}^0 = \prod_{a \in S_0^+} \Delta_{-a,k},$$

(5.1.26)
$$\nabla = \nabla_{S,k} = \Delta_{S,k} \Delta_{S,k}^0 .$$

If $i \in I$, $i \neq 0$, we have

$$\frac{s_i \Delta^0}{\Delta^0} = \frac{\Delta_{a_i} \Delta_{2a_i}}{\Delta_{-a_i} \Delta_{-2a_i}} = \frac{\Delta}{s_i \Delta}$$

and hence

(5.1.27) *∇ is W_0-symmetric.* □

(5.1.28) (i) When $S = S(R)$ with R reduced, we have

$$\nabla = \prod_{\alpha \in R} \frac{(e^\alpha; q)_\infty}{\left(q^{k(\alpha)} e^\alpha; q\right)_\infty}.$$

(ii) When $S = S(R)^\vee$ we have

$$\nabla = \prod_{\alpha \in R} \frac{\left(e^{\alpha^\vee}; q_\alpha\right)_\infty}{\left(q^{k^\vee(\alpha)} e^{\alpha^\vee}; q_\alpha\right)_\infty}$$

in the notation of (5.1.13).

(iii) When S is of type $(C_n^\vee, C_n)$ we have

$$\nabla = \nabla^{(1)} \nabla^{(2)}$$

where, in the notation of (5.1.14),

$$\nabla^{(1)} = \prod_{\alpha \in R_1} \frac{(e^{2\alpha}; q)_\infty}{\prod_{i=1}^4 (u_i e^\alpha; q)_\infty},$$

$$\nabla^{(2)} = \prod_{\alpha \in R_2} \frac{(e^\alpha; q)_\infty}{(q^{k_5} e^\alpha; q)_\infty}.$$

For $f, g \in A$ we define

(5.1.29) $$<f, g> = \frac{1}{|W_0|} \mathrm{ct}(f \bar{g} \nabla)$$

where $g \mapsto \bar{g}$ is the involution on A defined as follows: if $g = \sum g_\mu e^\mu$ then

(5.1.30) $$\bar{g} = \sum g_\mu e^{-\mu}.$$

Since $\nabla = \bar{\nabla}$, it follows that the scalar product (5.1.29) is *symmetric*:

(5.1.31) $$<f, g> = <g, f>.$$

The restrictions to $A_0 = A^{W_0}$ of the two scalar products are closely related. For each $w \in W_0$ let

(5.1.32) $$k(w) = \sum_{a \in S(w)} k(a)$$

where, as in Chapter 2, $S(w) = S^+ \cap w^{-1} S^-$, and let

(5.1.33) $$W_0(q^k) = \sum_{w \in W_0} q^{k(w)}.$$

(In multiplicative notation,

$$q^{k(w)} = \prod_{a \in S(w)} t_a = t_w$$

since $t_a t_{2a} = q^{k(a)+k(2a)}$ by (5.1.1).)

For $g = \sum g_\mu e^\mu \in A$, let

$$g^0 = \sum g_\mu^* e^\mu = \bar{g}^*. \tag{5.1.34}$$

Then for $f, g \in A_0$ we have

$$(f, g) = W_0(q^k) <f, g^0>. \tag{5.1.35}$$

Proof We have

$$\begin{aligned}(f, g) &= \mathrm{ct}(f g^* \Delta) \\ &= \frac{1}{|W_0|} \mathrm{ct}\left(f g^* \sum_{w \in W_0} w\Delta \right)\end{aligned}$$

and

$$\sum_{w \in W_0} w\Delta = \nabla \sum_{w \in W_0} w(\Delta^0)^{-1},$$

since $\nabla = \Delta\Delta^0$ is W_0-symmetric (5.1.27). Hence (5.1.35) follows from the identity

$$\sum_{w \in W_0} w(\Delta^0)^{-1} = W_0(q^k) \tag{5.1.36}$$

which is a well-known result ([M3], or (5.5.16) below). □

Finally, for $f, g \in A$ we define

$$<f, g>_1 = <f, g>/<1, 1>. \tag{5.1.37}$$

Then it follows from (5.1.35) that for $f, g \in A_0$ we have

$$(f, g)_1 = <f, g^0>_1 \tag{5.1.38}$$

and hence by (5.1.19)

$$<f^0, g>_1 = <f, g^0>_1^*. \tag{5.1.39}$$

We conclude this section with two results relating to the polynomial $W_0(q^k)$:

$$W_0(q^k) = \left(\Delta_{S,k}^0(-\rho_k')\right)^{-1}, \tag{5.1.40}$$

$$W_0(q^k) = W_0(q^{k'}). \tag{5.1.41}$$

Proof (5.1.40) follows from (5.1.36) by evaluating the left-hand side at $-\rho'_k$, which kills all the terms in the sum except that corresponding to $w = 1$ [M3].

As to (5.1.41), we may assume that S is of type $(C_n^\vee, C_n)$ (1.4.3), since $k' = k$ in all other cases. In that case,

$$k(w) = l_1(w)(k_1 + k_2) + l_2(w)k_5$$

where, in the notation of (5.1.14),

$$l_i(w) = \text{Card}\,\{\alpha \in R_i^+ : w\alpha \in R_i^-\}.$$

Since $k'_1 + k'_2 = k_1 + k_2$ and $k'_5 = k_5$, it follows that $k'(w) = k(w)$ for all $w \in W_0$, which gives (5.1.41). □

5.2 The polynomials E_λ

(5.2.1) *For each $\lambda \in L$ there is a unique element $E_\lambda \in A$ such that*
(i) $E_\lambda = e^\lambda +$ lower terms,
(ii) $(E_\lambda, e^\mu) = 0$ *for all* $\mu < \lambda$,
where "lower terms" means a K-linear combination of the e^μ, $\mu \in L$, such that $\mu < \lambda$.

Proof Let A_λ denote the finite-dimensional subspace of A spanned by the e^μ such that $\mu \leq \lambda$. By (5.1.21) the scalar product remains non-degenerate on restriction to A_λ. Hence the space of $f \in A_\lambda$ orthogonal to e^μ for each $\mu < \lambda$ is one-dimensional, i.e. the condition (ii) determines E_λ up to a scalar factor. Condition (i) then determines E_λ uniquely. □

Let $f \in A'$. Then we have

$$(f(Y)E_\lambda, e^\mu) = (E_\lambda, f^*(Y)e^\mu) = 0$$

if $\mu < \lambda$, by (5.1.24) and (4.6.12). It follows that $f(Y)E_\lambda$ is a scalar multiple of E_λ, namely

(5.2.2) $$f(Y)E_\lambda = f(-r_{k'}(\lambda))E_\lambda$$

by (4.6.12) again. Hence the E_λ form a K-basis of A that diagonalizes the action of $A'(Y)$ on A.

Moreover, the E_λ are pairwise orthogonal:

(5.2.3) $$(E_\lambda, E_\mu) = 0$$

if $\lambda \neq \mu$.

Proof Let $\nu' \in L'$. Then

$$q^{<\nu', r_{k'}(\lambda)>}(E_\lambda, E_\mu) = (Y^{-\nu'} E_\lambda, E_\mu) = (E_\lambda, Y^{\nu'} E_\mu)$$
$$= q^{<\nu', r_{k'}(\mu)>}(E_\lambda, E_\mu),$$

by (5.2.2) and (5.1.24). Assume first that $k'(\alpha^\vee) > 0$ for each $\alpha \in R$; if $\lambda \neq \mu$ we have $r_{k'}(\lambda) \neq r_{k'}(\mu)$ by (2.8.5), hence we can choose ν' so that $<\nu', r_{k'}(\lambda)> \neq <\nu', r_{k'}(\mu)>$, and we conclude that $(E_\lambda, E_\mu) = 0$ if all the labels $k'(\alpha^\vee)$ are positive.

Now the normalized scalar product $(E_\lambda, E_\mu)_1$ (5.1.18) is an element of K, that is to say a rational function in say r variables over $\mathbb{Q}$ (where $r \leq 6$). By the previous paragraph it vanishes on a non-empty open subset of $\mathbb{R}^r$, and hence identically. □

Dually, we have polynomials $E'_\mu \in A'$ for each $\mu \in L'$, satisfying

(5.2.1′) $$E'_\mu = e^\mu + \text{ lower terms},$$

(5.2.2′) $$f(Y)E'_\mu = f(-r'_k(\mu))E'_\mu$$

for each $f \in A$, and

(5.2.3′) $$(E'_\mu, E'_\nu)' = 0$$

if $\mu, \nu \in L'$ and $\mu \neq \nu$.

Next we have

(5.2.4) (Symmetry). *Let $\lambda \in L$, $\mu \in L'$. Then*

$$E_\lambda(r'_k(\mu))E'_\mu(-\rho_{k'}) = E_\lambda(-\rho'_k)E'_\mu(r_{k'}(\lambda)).$$

Proof From (4.7.13) we have

$$[E_\lambda, E'_\mu] = (E_\lambda(Y^{-1})E'_\mu)(-\rho_{k'})$$
$$= E_\lambda(r'_k(\mu))E'_\mu(-\rho_{k'})$$

by (5.2.2′). Hence the result follows from (4.7.14). □

We shall exploit (5.2.4) to calculate $E_\lambda(-\rho'_k)$ and the normalized scalar product $(E_\lambda, E_\lambda)_1$. When $t_a = 1$ for all $a \in S$, we have $\Delta = 1$, so that $E_\lambda = e^\lambda$ for each $\lambda \in L$; also $\rho'_k = 0$, so that $E_\lambda(-\rho'_k) = 1$. It follows that $E_\lambda(-\rho'_k)$ is not identically zero, so that we may define

(5.2.5) $$\tilde{E}_\lambda = E_\lambda / E_\lambda(-\rho'_k)$$

for $\lambda \in L$, and dually

(5.2.5′) $$\tilde{E}'_\mu = E'_\mu / E'_\mu(-\rho_{k'})$$

for $\mu \in L'$. Then (5.2.4) takes the form

(5.2.6) $$\tilde{E}_\lambda(r'_k(\mu)) = \tilde{E}'_\mu(r_{\kappa'}(\lambda)).$$

(5.2.7) *Let $\lambda \in L$. Then*

(i) $Y^{-\lambda} = \sum_w f_w(X) w^{-1}$,

(ii) $Y^\lambda = \sum_w w g_w(X)$

as operators on A', where $f_w, g_w \in A'[\mathbf{c}]$ and the summations are over $w \in W'$ such that $w \leq t(\lambda)$.

Proof Let $\lambda = \pi - \sigma$ where $\pi, \sigma \in L_{++}$, so that $Y^\lambda = T(t(\sigma))^{-1} T(t(\pi))$. Both (i) and (ii) now follow from (4.4.7). □

(5.2.8) *Let $\lambda, \mu \in L$. Then*

(i) $e^\lambda \tilde{E}_\mu = \sum_w f_w(r_{k'}(\mu)) \tilde{E}_{w\mu}$

with f_w as in (5.2.7) (i); *the summation is now over $w \in W'$ such that $w \leq t(\lambda)$ and $w(r_{k'}(\mu)) = r_{k'}(w\mu)$.*

(ii) $e^{-\lambda} \tilde{E}_\mu = \sum_w g_w(w^{-1} r_{k'}(\mu)) \tilde{E}_{w^{-1}\mu}$

with g_w as in (5.2.7) (ii); *the summation is now over $w \in W'$ such that $w \leq t(\lambda)$ and $w^{-1}(r_{k'}(\mu)) = r_{k'}(w^{-1}\mu)$.*

Proof Let $\nu \in L'$. Since

$$Y^{-\lambda} \tilde{E}'_\nu = q^{<\lambda, r'_k(\nu)>} \tilde{E}'_\nu$$

by (5.2.2′), it follows from (5.2.7) (i) that

$$q^{<\lambda, r'_k(\nu)>} \tilde{E}'_\nu = \sum_{w \leqslant t(\lambda)} f_w w^{-1} \tilde{E}'_\nu.$$

Now evaluate both sides at $r_{k'}(\mu)$ and use (5.2.6). We shall obtain

$$q^{<\lambda, r'_k(\nu)>} \tilde{E}_\mu(r'_k(\nu)) = \sum_w f_w(r_{k'}(\mu)) \tilde{E}_{w\mu}(r'_k(\nu))$$

summed over $w \in W'$ as stated above, since $w(r_{k'}(\mu)) = r_{k'}(w\mu)$ if $f_w(r_{k'}(\mu)) \neq 0$ by (4.5.3). Hence the two sides of (5.2.8) (i) agree at all points

$r'_k(\nu)$ where $\nu \in L'$, and therefore they are equal, by (4.5.8). The proof of (5.2.8) (ii) is similar. □

When $\mu = 0$, so that $\tilde{E}_\mu = E_\mu = 1$, (5.2.8) (i) gives

$$e^\lambda = \sum_w f_w(-\rho_{k'})\tilde{E}_{w(0)}$$

summed over $w \in W'$ such that $w \leq t(\lambda)$ and $w(-\rho_{k'}) = r_{k'}(w(0))$. Let us assume provisionally that the labelling k' is such that

(∗) $\rho_{k'}$ *is not fixed by any element* $w \neq 1$ *of* W'.

If $w(0) = \mu$ then $r_{k'}(w(0)) = u'(\mu)(-\rho_{k'})$ and it follows that $w = u'(\mu)$, so that

(5.2.9) $$e^\lambda = \sum_{\mu \leqslant \lambda} f_{u'(\mu)}(-\rho_{k'})\tilde{E}_\mu.$$

Hence, considering the coefficient of e^λ on the right-hand side, we have

(5.2.10) $$E_\lambda(-\rho'_k) = f_{u'(\lambda)}(-\rho_{k'}).$$

Next, when $\mu = \lambda$, (5.2.8) (ii) gives

$$e^{-\lambda}\tilde{E}_\lambda = \sum_w g_w(w^{-1}r_{k'}(\lambda))\tilde{E}_{w^{-1}\lambda},$$

summed over $w \in W'$ such that $w \leq t(\lambda)$ and $w^{-1}(r_{k'}(\lambda)) = r_{k'}(w^{-1}\lambda)$. In particular, if $w^{-1}\lambda = 0$ we have $w^{-1}(r_{k'}(\lambda)) = r_{k'}(0) = -\rho_{k'}$, so that $w = u'(\lambda)$. Hence the coefficient of $\tilde{E}_0 = 1$ in $e^{-\lambda}\tilde{E}_\lambda$ is equal to $g_{u'(\lambda)}(-\rho_{k'})$, and therefore by (5.2.9)

$$\begin{aligned}(\tilde{E}_\lambda, \tilde{E}_\lambda)_1 &= f_{u'(\lambda)}(-\rho_{k'})^{-1}(e^\lambda, \tilde{E}_\lambda)_1 \\ &= f_{u'(\lambda)}(-\rho_{k'})^{-1}(1, e^{-\lambda}\tilde{E}_\lambda)_1\end{aligned}$$

so that

(5.2.11) $$(\tilde{E}_\lambda, \tilde{E}_\lambda)_1 = g_{u'(\lambda)}(-\rho_{k'})^*/f_{u'(\lambda)}(-\rho_{k'}).$$

It remains to calculate f and g explicitly. From (5.2.7) and (4.4.8) we have

(i) $$T(v(\lambda)) \sum_{\substack{w \in W' \\ w(0)=\lambda}} f_w(X)w^{-1} = \boldsymbol{c}_{S',k'}(u'(\lambda))(X)u'(\lambda)^{-1}$$

and

(ii) $$\sum_{\substack{w \in W' \\ w(0)=\lambda}} wg_w(X) = \boldsymbol{c}_{S',k'}(u'(\lambda)^{-1})(X)u'(\lambda)T(v(\lambda)).$$

Since $S'(w^{-1}) = -wS'(w)$ (2.2.2), it follows that

$$w^{-1}(\boldsymbol{c}_{S',k'}(w^{-1})) = \boldsymbol{c}_{S',-k'}(w)$$

(since $\boldsymbol{c}(t, u; x^{-1}) = \boldsymbol{c}(t^{-1}, u^{-1}; x)$), so that

$$\boldsymbol{c}_{S',k'}(w^{-1}) = w(\boldsymbol{c}_{S',-k'}(w)). \tag{5.2.12}$$

Hence (ii) above may be rewritten as

(iii) $\displaystyle\sum_{\substack{w\in W' \\ w(0)=\lambda}} wg_w(X) = u'(\lambda)\boldsymbol{c}_{S',-k'}(u'(\lambda))(X)T(v(\lambda)).$

Now let

$$T(v(\lambda)) = \sum_{w\leq v(\lambda)} h_w(X)w.$$

By (4.5.6) we have $h_w(-\rho_{k'}) = 0$ if $w \neq 1$, and $h_1(-\rho_{k'}) = \tau_{v(\lambda)}$. Hence we obtain from (i)

$$f_{u'(\lambda)}(-\rho_{k'}) = \tau_{v(\lambda)}^{-1}\boldsymbol{c}_{S',\,k'}(u'(\lambda))(-\rho_{k'})$$

and from (iii)

$$g_{u'(\lambda)}(-\rho_{k'}) = \tau_{v(\lambda)}\boldsymbol{c}_{S',-k'}(u'(\lambda))(-\rho_{k'}).$$

Let

$$\varphi_\lambda^{\pm} = \boldsymbol{c}_{S',\pm k'}(u'(\lambda)^{-1}).$$

By (2.4.8) we have

$$\varphi_\lambda^{\pm} = \prod_{\substack{a'\in S_1'^{+} \\ a'(\lambda)<0}} \boldsymbol{c}_{a',\pm k'} \tag{5.2.13}$$

and from (5.2.12) we have

$$\boldsymbol{c}_{S',\pm k'}(u'(\lambda)) = u'(\lambda)^{-1}\varphi_\lambda^{\mp},$$

so that

$$f_{u'(\lambda)}(-\rho_{k'}) = \tau_{v(\lambda)}^{-1}\varphi_\lambda^{-}(r_{k'}(\lambda)),$$
$$g_{u'(\lambda)}(-\rho_{k'}) = \tau_{v(\lambda)}\varphi_\lambda^{+}(r_{k'}(\lambda)),$$

and hence finally

$$E_\lambda(-\rho_k') = \tau_{v(\lambda)}^{-1}\varphi_\lambda^{-}(r_{k'}(\lambda)), \tag{5.2.14}$$

$$(E_\lambda, E_\lambda)_1 = \varphi_\lambda^{+}(r_{k'}(\lambda))\varphi_\lambda^{-}(r_{k'}(\lambda)). \tag{5.2.15}$$

These relations have been derived under the restriction (∗) on the labelling k'. Since each of them asserts that two elements of K are equal, they are true identically.

5.3 The symmetric polynomials P_λ

For each $\lambda \in L_{++}$ let

$$m_\lambda = \sum_{\mu \in W_0\lambda} e^\mu,$$

the orbit-sum corresponding to λ.

(5.3.1) *For each $\lambda \in L_{++}$ there is a unique element $P_\lambda \in A_0$ such that*
(i) $P_\lambda = m_\lambda +$ lower terms,
(ii) $<P_\lambda, m_\mu> = 0$ *for all $\mu \in L_{++}$ such that $\mu < \lambda$.*

Here "lower terms" means a K-linear combination of the orbit-sums m_μ such that $\mu \in L_{++}$ and $\mu < \lambda$.
The proof is the same as that of (5.2.1).

Next, recall (5.1.33) that if $f \in A$, say

$$f = \sum_\lambda f_\lambda e^\lambda$$

with coefficients $f_\lambda \in K$, then

$$f^0 = \sum_\lambda f_\lambda^* e^\lambda.$$

(5.3.2) $P_\lambda^0 = P_\lambda$ *for each $\lambda \in L_{++}$.*

Proof By (5.1.37),

$$<P_\lambda^0, m_\mu>_1 = <P_\lambda, m_\mu>_1^* = 0$$

if $\mu < \lambda$. Hence P_λ^0 satisfies conditions (i) and (ii) of (5.3.1), and is therefore equal to P_λ. □

Let $f \in A_0'$. Then $f(Y)P_\lambda \in A_0$, by (4.3.18). Hence, by (5.1.24) and (5.1.37),

$$<f(Y)P_\lambda, m_\mu>_1 = <P_\lambda, (f^*(Y)m_\mu)^0>_1.$$

By (4.6.13), $(f^*(Y)m_\mu)^0$ is a linear combination of the m_ν such that $\nu \in L_{++}$ and $\nu \leq \mu$. It follows that $<f(Y)P_\lambda, m_\mu> = 0$ if $\mu < \lambda$, and hence that $f(Y)P_\lambda$ is a scalar multiple of P_λ. By (4.6.13) again, the scalar multiple is $f(-\lambda - \rho_{k'})$:

(5.3.3)
$$f(Y)P_\lambda = f(-\lambda - \rho_{k'})P_\lambda$$

for all $f \in A'_0$ and $\lambda \in L_{++}$.

From (5.3.3) it follows that

(5.3.4)
$$<P_\lambda, P_\mu> = 0$$

if $\lambda \neq \mu$. The proof is the same as that of (5.2.3).

Dually, we have symmetric polynomials $P'_{\mu'} \in A'_0$ for $\mu' \in L'_{++}$, satisfying the counterparts of (5.3.1)–(5.3.4).

Next, corresponding to (5.2.4), we have

(5.3.5) (Symmetry) *Let* $\lambda \in L_{++}$, $\mu' \in L'_{++}$. *Then*

$$P_\lambda(\mu' + \rho'_k)P'_{\mu'}(\rho_{k'}) = P_\lambda(\rho'_k)P'_{\mu'}(\lambda + \rho_{k'}).$$

Proof From (4.7.13) we have

$$[P_\lambda, P'_{\mu'}] = (P_\lambda(Y^{-1})P'_{\mu'})(-\rho_{k'})$$
$$= P_\lambda(\mu' + \rho'_k)P'_\mu(-\rho'_k)$$

by (5.3.3),

$$= P_\lambda(\mu' + \rho'_k)P'_{\mu'}(\rho_{k'})$$

since $-\rho_{k'} = w_0\rho_{k'}$, where w_0 is the longest element of W_0. Hence (5.3.5) follows from (4.7.14). □

As in §5.2, we shall exploit (5.3.5) to calculate $P_\lambda(\rho'_k)$ and the normalized scalar product $<P_\lambda, P_\lambda>_1$. The same argument as before shows that $P_\lambda(\rho'_k)$ is not identically zero, so that we may define

$$\tilde{P}_\lambda = P_\lambda / P_\lambda(\rho'_k)$$

for $\lambda \in L_{++}$, and dually

$$\tilde{P}'_{\mu'} = P'_{\mu'} / P'_{\mu'}(\rho_{k'})$$

for $\mu' \in L'_{++}$. Then (5.3.5) takes the form

$$\tilde{P}_\lambda(\mu' + \rho'_k) = \tilde{P}'_{\mu'}(\lambda + \rho_{k'}). \tag{5.3.6}$$

Let $\lambda \in L_{++}$. By (4.4.12) the restriction to A'_0 of the operator $m_\lambda(Y^{-1})$ is of the form

$$m_\lambda(Y^{-1})_0 = \sum_{\pi \in \Sigma(\lambda)} g_\pi(X)t(-\pi)$$

in which

$$g_{w\lambda} = w(\boldsymbol{c}'_\lambda) \tag{5.3.7}$$

for $w \in W_0$, where $\boldsymbol{c}'_\lambda = \boldsymbol{c}_{S',k'}(t(\lambda))$.

Let $\nu' \in L'_{++}$. Then $m_\lambda(Y^{-1})\tilde{P}'_{\nu'} = m_\lambda(\nu' + \rho'_k)\tilde{P}'_{\nu'}$, by (5.3.3), so that

$$m_\lambda(\nu' + \rho'_k)\tilde{P}'_{\nu'} = \sum_\pi g_\pi t(-\pi)\tilde{P}'_{\nu'}.$$

We shall evaluate both sides at $w_0(\mu + \rho_{k'}) = w_0\mu - \rho_{k'}$, where $\mu \in L_{++}$. We shall assume provisionally that

(*) $k'(a') \neq 0$ *for all* $a' \in S'$

so that we can apply (4.5.7), which shows that $g_\pi(w_0\mu - \rho_{k'}) = 0$ unless $\pi + w_0\mu$ is antidominant, i.e. unless $w_0\pi + \mu$ is dominant. We have then

$$\begin{aligned}(t(-\pi)\tilde{P}'_{\nu'})(w_0\mu - \rho_{k'}) &= \tilde{P}'_{\nu'}(\pi + w_0\mu - \rho_{k'}) \\ &= \tilde{P}'_{\nu'}(w_0\pi + \mu + \rho_{k'}) \\ &= \tilde{P}_{w_0\pi+\mu}(\nu' + \rho'_k)\end{aligned}$$

by (5.3.6), and therefore

$$m_\lambda(\nu' + \rho'_k)\tilde{P}_\mu(\nu' + \rho'_k) = \sum_\pi g_\pi(w_0\mu - \rho_{k'})\tilde{P}_{w_0\mu+\pi}(\nu' + \rho'_k)$$

for all dominant $\nu' \in L'$. Hence by (4.5.8) we have

$$m_\lambda\tilde{P}_\mu = \sum g_\pi(w_0\mu - \rho_{k'})\tilde{P}_{\mu+w_0\pi} \tag{5.3.8}$$

summed over $\pi \in \Sigma(\lambda)$ such that $\mu + w_0\pi$ is dominant.

In particular, when $\mu = 0$, (5.3.8) expresses m_λ as a linear combination of the $\tilde{P}_\pi$, $\pi \leq \lambda$:

$$m_\lambda = \sum_{\pi \leq \lambda} g_{w_0\pi}(-\rho_{k'})\tilde{P}_\pi$$

in which the coefficient of $\tilde{P}_\lambda$ is

$$g_{w_0\lambda}(-\rho_{k'}) = (w_0 \boldsymbol{c}'_\lambda)(-\rho_{k'}) = \boldsymbol{c}'_\lambda(\rho_{k'})$$

by (5.3.7). It follows that

$$P_\lambda(\rho'_k) = \boldsymbol{c}'_\lambda(\rho_{k'}). \tag{5.3.9}$$

Next, let $\bar{\lambda} = -w_0\lambda$, and replace (λ, μ) in (5.3.8) by $(\bar{\lambda}, \lambda)$. Since $m_{\bar{\lambda}} = \bar{m}_\lambda$, it follows that

$$\bar{m}_\lambda \tilde{P}_\lambda = \sum_\pi g_\pi(-\bar{\lambda} - \rho_{k'})\tilde{P}_{\lambda + w_0\pi}$$

in which the coefficient of $\tilde{P}_0 = 1$ is

$$g_{\bar{\lambda}}(-\bar{\lambda} - \rho_{k'}) = \boldsymbol{c}'_{\bar{\lambda}}(-\bar{\lambda} - \rho_{k'}) = \boldsymbol{c}'_\lambda(-\lambda - \rho_{k'})$$

since $\bar{\rho}_{k'} = \rho_{k'}$. Hence

$$\begin{aligned} <\tilde{P}_\lambda, \tilde{P}_\lambda>_1 &= \boldsymbol{c}'_\lambda(\rho_{k'})^{-1} <m_\lambda, \tilde{P}_\lambda>_1 \\ &= \boldsymbol{c}'_\lambda(\rho_{k'})^{-1} <1, \bar{m}_\lambda \tilde{P}_\lambda>_1 \\ &= \boldsymbol{c}'_\lambda(-\lambda - \rho_{k'})/\boldsymbol{c}'_\lambda(\rho_{k'}) \end{aligned}$$

and therefore, by (5.3.9),

$$<P_\lambda, P_\lambda>_1 = \boldsymbol{c}'_\lambda(-\lambda - \rho_{k'})\, \boldsymbol{c}'_\lambda(\rho_{k'}). \tag{5.3.10}$$

We have derived (5.3.9) and (5.3.10) under the restriction (∗) on the labelling k'. For the same reason as in §5.2, they are identically true.

The formulas (5.3.9) and (5.3.10) can be restated, as follows.
Let

$$\Delta^+_{S,k} = \prod_{\substack{a \in S^+ \\ Da > 0}} \Delta_a, \quad \Delta^-_{S,k} = \prod_{\substack{a \in S^+ \\ Da < 0}} \Delta_a \tag{5.3.11}$$

and define $\Delta^+_{S',k'}$, analogously. Then

$$P_\lambda(\rho'_k) = q^{-<\lambda, \rho'_k>} \Delta^+_{S',k'}(\lambda + \rho_{k'})/\Delta^+_{S',k'}(\rho_{k'}), \tag{5.3.12}$$

$$<P_\lambda, P_\lambda>_1 = \frac{\Delta^+_{S',k'}(\lambda + \rho_{k'})\Delta^-_{S',-k'}(-\lambda - \rho_{k'})}{\Delta^+_{S',k'}(\rho_{k'})\Delta^-_{S',-k'}(-\rho_{k'})}. \tag{5.3.13}$$

Proof We shall verify these formulas when $S = S(R)$ (1.4.1); the other cases are similar. We have $S' = S(R^\vee)$ and $k'(\alpha^\vee) = k(\alpha)$, so that

$$\Delta^+_{S',k'} = \prod_{\alpha\in R^+} \frac{(e^{\alpha^\vee};q)_\infty}{\left(q^{k(\alpha)}e^{\alpha^\vee};q\right)_\infty}.$$

Since λ is dominant, we have

$$S'(t(\lambda)) = \{\alpha^\vee + rc : \alpha\in R^+ \text{ and } 0\le r < <\lambda,\alpha^\vee>\}.$$

Hence

$$c'_\lambda = c_{S',k'}(t(\lambda)) = \prod_{\alpha\in R^+}\prod_{r=0}^{<\lambda,\alpha^\vee>-1} q^{-k(\alpha)/2}\frac{1-q^{k(\alpha)+r}e^{\alpha^\vee}}{1-q^r e^{\alpha^\vee}}$$

and therefore by (5.3.9)

$$P_\lambda(\rho'_k) = q^{-<\lambda,\rho'_k>}\prod_{\alpha\in R^+}\frac{\left(q^{k(\alpha)+<\rho_{k'},\alpha^\vee>};q\right)_{<\lambda,\alpha^\vee>}}{\left(q^{<\rho_{k'},\alpha^\vee>};q\right)_{<\lambda,\alpha^\vee>}}$$

which gives (5.3.12).

Next, we have

$$\begin{aligned}
c'_\lambda(-\lambda-\rho_{k'}) &= \prod_{\alpha\in R^+}\prod_{r=0}^{<\lambda,\alpha^\vee>-1} q^{-k(\alpha)/2}\frac{1-q^{k(\alpha)+r-<\lambda+\rho_{k'},\alpha^\vee>}}{1-q^{r-<\lambda+\rho_{k'},\alpha^\vee>}}\\
&= \prod_{\alpha\in R^+}\prod_{r=0}^{<\lambda,\alpha^\vee>-1} q^{k(\alpha)/2}\frac{1-q^{<\lambda+\rho_{k'},\alpha^\vee>-r-k(\alpha)}}{1-q^{<\lambda+\rho_{k'},\alpha^\vee>-r}}\\
&= \prod_{\alpha\in R^+}\prod_{r'=0}^{<\lambda,\alpha^\vee>-1} q^{k(\alpha)/2}\frac{1-q^{<\rho_{k'},\alpha^\vee>+1+r'-k(\alpha)}}{1-q^{<\rho_{k'},\alpha^\vee>+1+r'}}
\end{aligned}$$

(where $r' = <\lambda,\alpha^\vee> - 1 - r$ in the last product above). Since

$$\Delta^-_{S',-k'} = \prod_{\alpha\in R^+}\frac{(qe^{-\alpha^\vee};q)_\infty}{\left(q^{1-k(\alpha)}e^{-\alpha^\vee};q\right)_\infty}$$

it follows that

$$c'_\lambda(-\lambda-\rho_{k'}) = q^{<\lambda,\rho'_k>}\Delta^-_{S',-k'}(-\lambda-\rho_{k'})/\Delta^-_{S',-k'}(-\rho_{k'})$$

which together with (5.3.12) gives (5.3.13). □

To conclude this section we shall consider some special cases.

(5.3.14) When $k(a) = 0$ for all $a\in S$, we have $\nabla = 1$ and P_λ is the orbit-sum m_λ, for all $\lambda\in L_{++}$.

(5.3.15) Suppose that $S = S(R)$, with R reduced, and that $k(\alpha) = 1$ for all $\alpha \in R$. Then

$$\nabla = \prod_{\alpha \in R}(1 - e^\alpha) = \prod_{\alpha \in R}\left(e^{\alpha/2} - e^{-\alpha/2}\right) = \delta\bar{\delta},$$

where

$$\delta = \delta_R = \prod_{\alpha \in R^+}\left(e^{\alpha/2} - e^{-\alpha/2}\right) = \sum_{w \in W_0}(-1)^{l(w)}e^{w\rho}$$

by Weyl's denominator formula, where

$$\rho = \frac{1}{2}\sum_{\alpha \in R^+}\alpha.$$

For $\lambda \in L_{++}$, let

$$\chi_\lambda = \chi_{R,\lambda} = \delta^{-1}\sum_{w \in W_0}(-1)^{l(w)}e^{w(\lambda+\rho)} \in A_0.$$

Then

$$\chi_\lambda = m_\lambda + \text{lower terms},$$

and

$$<\chi_\lambda, \chi_\mu> = \frac{1}{|W_0|}\text{ ct }(\chi_\lambda\delta \cdot \overline{\chi_\mu\delta})$$

is zero if $\lambda \neq \mu$, and is equal to 1 if $\lambda = \mu$. It follows that $P_\lambda = \chi_{R,\lambda}$ in this case.

When $S = S(R)^\vee$ and $k^\vee(\alpha) = 1$ for all $\alpha \in R$, in the notation of (5.1.13), the conclusion is the same: $P_\lambda = \chi_{R^\vee,\lambda}$.

Finally, when S is of type $(C_n^\vee, C_n)$ and $k_1 = k_2 = k_5 = 1$, $k_3 = k_4 = 0$, we have $\nabla = \delta_R\bar{\delta}_R$ where R is of type C_n, and consequently $P_\lambda = \chi_{R,\lambda}$.

(5.3.16) Consider next the case where $q \to 0$, the t_a being arbitrary. Then

$$\nabla = \prod_{a \in S_0}\frac{1 - t_{2a}^{1/2}e^a}{1 - t_a t_{2a}^{1/2}e^a}$$

where $S_0 = \{a \in S : a(0) = 0\}$. In this case there is an explicit formula for P_λ, namely

$$P_\lambda = W_{0\lambda}(t)^{-1} \sum_{w \in W_0} w \left(e^\lambda \prod_{a \in S_0^+} \frac{1 - t_a t_{2a}^{1/2} e^{-a}}{1 - t_{2a}^{1/2} e^{-a}} \right),$$

where $W_{0\lambda}$ is the subgroup of W_0 that fixes λ, and

$$W_{0\lambda}(t) = \sum_{w \in W_{o\lambda}} t_w$$

with t_w as defined in §5.1. Moreover

$$<P_\lambda, P_\lambda> = W_{0\lambda}(t)^{-1}$$

in this case. (For details see [M5], §10.)

(5.3.17) Finally, when $S = S(R)$ with R of type A_{n-1}, the P_λ are essentially the symmetric polynomials $P_\lambda(x; q, t)$ of [M6], Ch. 5.

When S is of type $(C_n^\vee, C_n)$, the P_λ are Koornwinder's orthogonal polynomials [K3]. In particular, when $n = 1$ they are the orthogonal polynomials (in one variable) of Askey and Wilson [A2].

5.4 The $\mathfrak{H}$-modules A_λ

As in §5.3, we shall assume provisionally that

(∗) $k'(a') \neq 0$ *for all* $a' \in S'$.

(5.4.1) *Let* $f \in A$, $f \neq 0$ *be a simultaneous eigenfunction of the operators* $Y^{\lambda'}(\lambda' \in L')$, *so that* $Y^{\lambda'} f = g_{\lambda'} f$ *for all* $\lambda' \in L'$ *and scalars* $g_{\lambda'}$. *Then* f *is a scalar multiple of* E_μ *for some* $\mu \in L$, *and* $g_{\lambda'} = q^{-<\lambda', r_{k'}(\mu)>}$ *for all* $\lambda' \in L'$.

Proof Since the E_μ form a K-basis of A we have

$$f = \sum_{\mu \in L} f_\mu E_\mu$$

with coefficients $f_\mu \in K$. Hence

$$Y^{\lambda'} f = \sum_\mu f_\mu q^{-<\lambda', r_{k'}(\mu)>} E_\mu$$

by (5.2.2). But also

$$Y^{\lambda'} f = \sum_\mu g_{\lambda'} f_\mu E_\mu$$

and therefore

$$f_\mu\big(g_{\lambda'} - q^{-<\lambda', r_{k'}(\mu)>}\big) = 0$$

for all $\lambda' \in L'$ and $\mu \in L$. Since $f \neq 0$ we have $f_\mu \neq 0$ for some $\mu \in L$, and therefore $g_{\lambda'} = q^{-<\lambda', r_{k'}(\mu)>}$ for all $\lambda' \in L'$. If $\nu \neq \mu$ then $r_{k'}(\nu) \neq r_{k'}(\mu)$ by (2.8.5), and therefore $f_\nu = 0$; consequently f is a scalar multiple of E_μ. □

(5.4.2) *Let* $\lambda \in L, i \in I, i \neq 0,$ *and let*

$$\boldsymbol{b}'_i = \boldsymbol{b}(\tau_i, \upsilon_i; e^{a'_i})$$

where (as in (4.2.4)) $\upsilon_i = \tau_i$ *or* τ_0 *according as* $<L, \alpha_i^\vee> = \mathbb{Z}$ *or* $2\mathbb{Z}$*. Then*

$$E = T_i E_\lambda - \boldsymbol{b}'_i(r_{k'}(\lambda)) E_\lambda$$

is a scalar multiple of $E_{s_i\lambda}$*, and is zero if* $\lambda = s_i\lambda$*.*

Proof Let

$$F_i = T_i - \boldsymbol{b}'_i(Y^{-1})$$

as operator on A, so that $E = F_i E_\lambda$ by (5.2.2). By (4.2.4) we have

$$Y^{\lambda'} F_i = F_i Y^{s_i\lambda'}$$

for $\lambda' \in L'$, hence

$$Y^{\lambda'} E = Y^{\lambda'} F_i E_\lambda = F_i Y^{s_i\lambda'} E_\lambda = q^{-<\lambda', s_i r_{k'}(\lambda)>} E.$$

If $\lambda \neq s_i\lambda$ then $s_i r_{k'}(\lambda) = r_{k'}(s_i\lambda)$ by (2.8.4), and hence E is a scalar multiple of $E_{s_i\lambda}$ by (5.4.1). If $\lambda = s_i\lambda$ then $s_i(r_{k'}(\lambda)) \notin r_{k'}(L)$ by (2.8.6), and hence $E = 0$ by (5.4.1). □

(5.4.3) *Let* $\lambda \in L, i \neq 0$*. If* $<\lambda, \alpha'_i> > 0$ *then*

$$T_i E_\lambda = \tau_i^{-1} E_{s_i\lambda} + \boldsymbol{b}'_i(r_{k'}(\lambda)) E_\lambda.$$

Proof Since $<\lambda, \alpha'_i> > 0$, we have $s_i\lambda > \lambda$ (2.7.9) and hence it follows from (4.3.21) that

(1) $$T_i e^\lambda = \tau_i^{-1} e^{s_i\lambda} + \text{lower terms}.$$

On the other hand, by (5.4.2),

$$T_i E_\lambda = u E_{s_i\lambda} + \boldsymbol{b}'_i(r_{k'}(\lambda))E_\lambda$$

for some $u \in K$, and u is the coefficient of $e^{s_i\lambda}$ in $T_i E_\lambda$. If $\mu < \lambda$ then $T_i e^\mu$ contains only e^μ and $e^{s_i\mu}$ from the W_0-orbit of e^μ, and since $s_i\lambda > \lambda > \mu$ we have $\mu \neq s_i\lambda$ and $s_i\mu \neq s_i\lambda$. Hence it follows from (1) that $u = \tau_i^{-1}$. □

(5.4.4) *If $\lambda = s_i\lambda$ then $E_\lambda = s_i E_\lambda$.*

Proof If $\lambda = s_i\lambda$ we have $\boldsymbol{b}'_i(r_{k'}(\lambda)) = \tau_i$, by (4.2.3) (i) and (4.5.2). Hence $T_i E_\lambda = \tau_i E_\lambda$ by (5.4.2) and therefore $E_\lambda = s_i E_\lambda$ by (4.3.12). □

Let $\lambda \in L_{++}$ and let A_λ denote the K-span of the E_μ for $\mu \in W_0\lambda$.

(5.4.5) (i) *A_λ is an irreducible $\mathfrak{H}$-submodule of A.*
(ii) $A_\lambda = \mathfrak{H}_0 E_\lambda$.

Proof (i) By (5.2.2) and (5.4.2), A_λ is stable under the operators $Y^{\lambda'}$ ($\lambda' \in L'$) and T_i ($i \in I$, $i \neq 0$), hence is an $\mathfrak{H}$-submodule of A.

Let M be a nonzero $\mathfrak{H}$-submodule of A_λ and let

$$E = \sum_{i=1}^{r} a_i E_{\mu_i}$$

be a nonzero element of M, in which the μ_i are distinct elements of the orbit $W_0\lambda$, the coefficients a_i are $\neq 0$, and r is as small as possible. Then

$$Y^{-\lambda'} E = \sum_{i=1}^{r} a_i q^{<\lambda', r_{k'}(\mu_i)>} E_{\mu_i} \in M$$

for all $\lambda' \in L'$, and hence if $r > 1$

$$q^{<\lambda', r_{k'}(\mu_1)>} E - Y^{-\lambda'} E = \sum_{i=2}^{r} a_i \left(q^{<\lambda', r_{k'}(\mu_1)>} - q^{<\lambda', r_{k'}(\mu_i)>}\right) E_{\mu_i}$$

is a nonzero element of M, contradicting our choice of r. We therefore conclude that $r = 1$, i.e. that $E_\mu \in M$ for some $\mu \in W_0\lambda$. But then it follows from (5.4.2) that $E_{s_i\mu} \in M$ for all $i \neq 0$, and hence that $E_\nu \in M$ for all $\nu \in W_0\lambda$, so that $M = A_\lambda$. Hence A_λ is irreducible as an $\mathfrak{H}$-module.

(ii) Let $w \in W_0$ and let $\mu = w\lambda$. It follows from (5.4.2) and (5.4.3) that $T(w)E_\lambda$ is of the form

$$T(w)E_\lambda = \sum_{\nu \le \mu} a_{\mu\nu} E_\nu$$

with $a_{\mu\mu} = \tau_w^{-1} \neq 0$. Hence the $T(w)E_\lambda$, $w \in W_0$, span A_λ. □

(5.4.6) *If $\lambda \in L_{++}$ is regular, then A_λ is a free $\mathfrak{H}_0$-module of rank* 1, *generated by E_λ.*

This follows from (5.4.5) (ii), since $\dim A_\lambda = |W_0| = \dim \mathfrak{H}_0$. □

Now let $w \in W_0$, let $w = s_{i_1} \cdots s_{i_p}$ be a reduced expression, and let $\beta_r = s_{i_p} \cdots s_{i_{r+1}}(\alpha_{i_r})$ $(1 \le r \le p)$, so that $\{\beta_1, \ldots, \beta_p\} = S(w)$. Also, for each $\alpha \in R$, let $\boldsymbol{b}'_\alpha = w\boldsymbol{b}'_i$ if $\alpha = w\alpha_i$.

(5.4.7) *Let $w \in W_0$, $x \in r_{k'}(L)$. Then*

$$F_w(x) = \left(T_{i_1} - \boldsymbol{b}'_{\beta_1}(x)\right) \cdots \left(T_{i_p} - \boldsymbol{b}'_{\beta_p}(x)\right)$$

is independent of the reduced expression $s_{i_1} \cdots s_{i_p}$ of w.

Proof Let $\lambda \in L$ be regular dominant and let $\lambda_r = s_{i_{r+1}} \cdots s_{i_p}\lambda$ for $0 \le r \le p$. Then $<\lambda_r, \alpha_{i_r}^\vee> = <\lambda, \beta_r^\vee> > 0$ and therefore by (5.4.3)

$$\begin{aligned}\tau_{i_r}^{-1} E_{\lambda_{r-1}} &= \left(T_{i_r} - \boldsymbol{b}'_{i_r}(r_{k'}(\lambda_r))\right) E_{\lambda_r} \\ &= \left(T_{i_r} - \boldsymbol{b}'_{\beta_r}(r_{k'}(\lambda))\right) E_{\lambda_r}\end{aligned} \tag{1}$$

since

$$<r_{k'}(\lambda_r), \alpha_{i_r}^\vee> = <s_{i_{r+1}} \cdots s_{i_p}(r_{k'}(\lambda)), \alpha_{i_r}^\vee> = <r_{k'}(\lambda), \beta_r^\vee>.$$

Let

$$F_{i_1 \cdots i_p}(x) = \left(T_{i_1} - \boldsymbol{b}'_{\beta_1}(x)\right) \cdots \left(T_{i_p} - \boldsymbol{b}'_{\beta_p}(x)\right).$$

Then it follows from (1) that

$$F_{i_1 \cdots i_p}(r_{k'}(\lambda)) E_\lambda = \tau_w^{-1} E_{w\lambda}.$$

If $w = s_{j_1} \cdots s_{j_p}$ is another reduced expression, then likewise

$$F_{j_1 \cdots j_p}(r_{k'}(\lambda)) E_\lambda = \tau_w^{-1} E_{w\lambda}$$

and therefore $F_{i_1\cdots i_p}(r_{k'}(\lambda)) = F_{j_1\cdots j_p}(r_{k'}(\lambda))$ by (5.4.6). So (5.4.7) is true whenever $x = r_{k'}(\lambda) = \lambda + \rho_{k'}$ with $\lambda \in L$ regular dominant, and hence for all $x \in r_{k'}(L)$ by (4.5.8). □

Finally, the results in this section have been obtained under the restriction $(*)$ on the labelling k'. For the same reason as before, this restriction can now be lifted.

5.5 Symmetrizers

From (5.1.23), the adjoint of s_i is

(5.5.1) $$s_i^* = \frac{\boldsymbol{c}_i(X)}{\boldsymbol{c}_i(X^{-1})} s_i.$$

Let ε be a linear character of W_0, so that $\varepsilon(s_i) = \pm 1$ for each $i \neq 0$, and $\varepsilon(s_i) = \varepsilon(s_j)$ if s_i and s_j are conjugate in W_0. (If R is simply-laced, there are just two possibilities for ε, namely the trivial character and the sign character. In the other cases there are four possibilities for ε.) Define

$$s_i^\varepsilon = \begin{cases} s_i & \text{if } \varepsilon(s_i) = 1, \\ s_i^* & \text{if } \varepsilon(s_i) = -1. \end{cases}$$

(5.5.2) *Let $w \in W_0$ and let $w = s_{i_1} \cdots s_{i_p}$ be a reduced expression for w. Then*

$$w^{(\varepsilon)} = s_{i_1}^{(\varepsilon)} \cdots s_{i_p}^{(\varepsilon)}$$

depends only on w (and ε) and not on the reduced expression chosen. Hence $w \mapsto w^{(\varepsilon)}$ is an isomorphism of W_0 onto a subgroup $W_0^{(\varepsilon)}$ of Aut (A).

Proof This is a matter of checking the braid relations for the $s_i^{(\varepsilon)}$. Hence we may assume that R has rank 2, with basis $\{\alpha_i, \alpha_j\}$. One checks easily, using (5.5.1), that

$$\left(s_i^{(\varepsilon)} s_j^{(\varepsilon)}\right)^m = (s_i s_j)^m = 1,$$

where $m = \text{Card}(R^+)$. (Here the nature of the factors $\boldsymbol{c}_i(X)/\boldsymbol{c}_i(X^{-1})$ in (5.5.1) is immaterial: they could be replaced by any $f_i(X)$ such that $f_i(X)s_i = s_i f_i(X)^{-1}$.) □

Since $s_i X^\mu = X^{s_i\mu} s_i$ and hence also $s_i^* X^\mu = X^{s_i\mu} s_i^*$, it follows that

(5.5.3) $$w^{(\varepsilon)} X^\mu = X^{w\mu} w^{(\varepsilon)}$$

for all $w \in W_0$ and $\mu \in L$.

Next, let

(5.5.4) $$\tau_i^{(\varepsilon)} = \begin{cases} \tau_i & \text{if } \varepsilon(s_i) = 1, \\ -\tau_i^{-1} & \text{if } \varepsilon(s_i) = -1, \end{cases}$$

and for $w \in W_0$ let

(5.5.5) $$\tau_w^{(\varepsilon)} = \tau_{i_1}^{(\varepsilon)} \cdots \tau_{i_p}^{(\varepsilon)}$$

where as above $w = s_{i_1} \cdots s_{i_p}$ is a reduced expression. From (5.5.2) it follows that $\tau_w^{(\varepsilon)}$ is independent of the reduced expression chosen.

We now define the *ε-symmetrizer* U_ε by

(5.5.6) $$U_\varepsilon = \left(\tau_{w_0}^{(\varepsilon)}\right)^{-1} \sum_{w \in W_0} \tau_w^{(\varepsilon)} T(w),$$

where as usual w_0 is the longest element of W_0. When ε is the trivial character, we write U^+ for U_ε, so that

(5.5.7) $$U^+ = \tau_{w_0}^{-1} \sum_{w \in W_0} \tau_w T(w),$$

and when ε is the sign character we write U^- for U_ε, so that

(5.5.8) $$U^- = (-1)^{l(w_0)} \tau_{w_0} \sum_{w \in W_0} (-1)^{l(w)} \tau_w^{-1} T(w).$$

(5.5.9) *We have*

$$\left(T_i - \tau_i^{(\varepsilon)}\right) U_\varepsilon = U_\varepsilon \left(T_i - \tau_i^{(\varepsilon)}\right) = 0$$

for all $i \in I, i \neq 0$.

Proof Let $w \in W_0$. If $l(s_i w) > l(w)$ then

$$\left(T_i - \tau_i^{(\varepsilon)}\right) \tau_w^{(\varepsilon)} T(w) = \tau_w^{(\varepsilon)} T(s_i w) - \tau_{s_i w}^{(\varepsilon)} T(w).$$

If on the other hand $l(s_i w) < l(w)$, then

$$T(s_i w) = T_i^{-1} T(w) = \left(T_i - \tau_i^{(\varepsilon)} + \left(\tau_i^{(\varepsilon)}\right)^{-1}\right) T(w)$$

so that again

$$\left(T_i - \tau_i^{(\varepsilon)}\right)\tau_w^{(\varepsilon)}T(w) = \tau_w^{(\varepsilon)}T(s_i w) - \tau_{s_i w}^{(\varepsilon)}T(w).$$

Hence

$$\left(T_i - \tau_i^{(\varepsilon)}\right)U_\varepsilon = \left(\tau_{w_0}^{(\varepsilon)}\right)^{-1}\sum_{w\in W_0}\left(\tau_w^{(\varepsilon)}T(s_i w) - \tau_{s_i w}^{(\varepsilon)}T(w)\right) = 0.$$

Likewise $U_\varepsilon(T_i - \tau_i^{(\varepsilon)}) = 0$. □

Conversely:

(5.5.10) (i) *Let* $h \in A(X)\,\mathfrak{H}_0$ *be such that* $h(T_i - \tau_i^{(\varepsilon)}) = 0$ *for all* $i \neq 0$ *in* I. *Then* $h = f(X)U_\varepsilon$ *for some* $f \in A$.
(ii) *Let* $h \in A(X)\,\mathfrak{H}_0$ *be such that* $(T_i - \tau_i^{(\varepsilon)})h = 0$ *for all* $i \neq 0$. *Then* $h = U_\varepsilon f(X)$ *for some* $f \in A$.

Proof We shall prove (i); the proof of (ii) is analogous. We have $T(w)T_i = T(ws_i)$ if $l(ws_i) > l(w)$, and

$$T(w)T_i = T(ws_i) + \left(\tau_i^{(\varepsilon)} - \left(\tau_i^{(\varepsilon)}\right)^{-1}\right)T(w)$$

if $l(ws_i) < l(w)$. Let $h = \sum_{w\in W_0} f_w(X)T(w)$. Then

$$hT_i = \sum_{w\in W_0} f_w(X)T(ws_i) + \left(\tau_i^{(\varepsilon)} - \left(\tau_i^{(\varepsilon)}\right)^{-1}\right)\sum_{w} f_w(X)T(w),$$

where the second sum is over $w \in W_0$ such that $l(ws_i) < l(w)$. Since $hT_i = \tau_i^{(\varepsilon)}h$ it follows that $\tau_i^{(\varepsilon)}f_w = f_{ws_i}$ if $l(ws_i) > l(w)$, and hence that $f_w = \tau_w^{(\varepsilon)}f_1$ for all $w \in W_0$.

Consequently $h = \tau_{w_0}^{(\varepsilon)}f_1(X)U_\varepsilon$. □

Now let

(5.5.11) $$\rho_{\varepsilon k'} = \frac{1}{2}\sum_{\alpha\in R^+}\varepsilon(s_\alpha)k'(\alpha^\vee)\alpha.$$

Then we have

(5.5.12) $$U_\varepsilon = F_{w_0}(\rho_{\varepsilon k'}).$$

where F_w is defined by (5.4.7).

Proof Let $i \in I$, $i \neq 0$. Then there exists a reduced expression for w_0 ending with s_i. From (4.5.2) we have $\boldsymbol{c}_i(-\rho'_k) = 0$ and hence, by (4.2.3) (i), $\boldsymbol{b}_i(\rho'_k) = -\tau_i^{-1}$. Dually, therefore,

$$\boldsymbol{b}'_i(\rho_{\varepsilon k'}) = -\big(\tau_i^{(\varepsilon)}\big)^{-1},$$

Hence $F_{w_0}(\rho_{\varepsilon k'})$ is divisible on the right by $T_i + (\tau_i^{(\varepsilon)})^{-1}$, and therefore $F_{w_0}(\rho_{\varepsilon k'})$ $(T_i - \tau_i^{(\varepsilon)}) = 0$. It now follows from (5.5.10) that $F_{w_0}(\rho_{\varepsilon k'})$ is a scalar multiple of U_ε. Since the coefficient of $T(w_0)$ in each of U_ε and F_{w_0} is equal to 1, the result follows, □

Next, let

(5.5.13) $$V_\varepsilon = \varepsilon(w_0) \sum_{w \in W} \varepsilon(w) w^{(\varepsilon)}.$$

Then we have

(5.5.14) $$U_\varepsilon = V_\varepsilon \boldsymbol{c}_+\big(X^{-\varepsilon}\big)$$

where

(5.5.15) $$\boldsymbol{c}_+(X^{-\varepsilon}) = \prod_{a \in S_0^+} \boldsymbol{c}_a\big(X^{-\varepsilon(s_a)}\big)$$

in which S_0 *is the reduced root system with basis* $\{a_i : i \in I_0\}$.

Proof From (4.3.12), (4.3.13) and (5.5.1) we have, for $i \in I_0$,

$$T_i + \big(\tau_i^{(\varepsilon)}\big)^{-1} = \big(s_i^{(\varepsilon)} + \varepsilon(s_i)\big)\boldsymbol{c}_i\big(X^{-\varepsilon(s_i)}\big)$$

(which is precisely (5.5.15) in rank 1). Hence

$$\begin{aligned} V_\varepsilon \boldsymbol{c}_+(X^{-\varepsilon})\big(T_i + \big(\tau_i^{(\varepsilon)}\big)^{-1}\big) &= V_\varepsilon \boldsymbol{c}_+(X^{-\varepsilon})\big(s_i^{(\varepsilon)} + \varepsilon(s_i)\big)\boldsymbol{c}_i\big(X^{-\varepsilon(s_i)}\big) \\ &= \varepsilon(s_i) V_\varepsilon \boldsymbol{c}_+(X^{-\varepsilon})\boldsymbol{c}_i\big(X^{-\varepsilon(s_i)}\big) + V_\varepsilon s_i^{(\varepsilon)} \boldsymbol{c}_+(X^{-\varepsilon})\boldsymbol{c}_i\big(X^{\varepsilon(s_i)}\big) \\ &= \varepsilon(s_i) V_\varepsilon \boldsymbol{c}_+(X^{-\varepsilon})\big(\tau_i + \tau_i^{-1}\big), \end{aligned}$$

by (4.2.3) (ii), since $V_\varepsilon s_i^{(\varepsilon)} = \varepsilon(s_i) V_\varepsilon$. It follows that

$$V_\varepsilon \boldsymbol{c}_+\big(X^{-\varepsilon})(T_i - \tau_i^{(\varepsilon)}\big) = 0$$

for all $i \in I_0$, and hence by (5.5.10) that

$$V_\varepsilon \boldsymbol{c}_+(X^{-\varepsilon}) = f(X) U_\varepsilon$$

for some $f \in A$. It remains to show that $f = 1$, which we do by considering the coefficient of $w_0^{(\varepsilon)}$ on either side. Since by (4.3.14)

$$(1) \qquad T_i = \boldsymbol{b}_i(X) + s_i^{(\varepsilon)} \boldsymbol{c}_i(X^{-\varepsilon(s_i)}),$$

the coefficient of $w_0^{(\varepsilon)}$ in U_ε comes only from $T(w_0)$; also from (1) it follows that

$$T(w_0) = w_0^{(\varepsilon)} \boldsymbol{c}_+(X^{-\varepsilon}) + \text{lower terms}.$$

Hence $f = 1$ as required. □

In particular, let us take ε to be the trivial character of W_0, and evaluate both sides of (5.5.14) at 1_A, the identity element of A. We shall obtain

$$(5.5.16) \qquad W_0(t) = \sum_{w \in W_0} w(\Delta^0)^{-1}$$

in the notation of §5.1.

(5.5.17) (i) $T(w)U_\varepsilon = U_\varepsilon T(w) = \tau_w^{(\varepsilon)} U_\varepsilon$ *for all* $w \in W_0$.
(ii) $U_\varepsilon^2 = (\tau_{w_0}^{(\varepsilon)})^{-1} W_0(t^{(\varepsilon)}) U_\varepsilon$, *where* $W_0(t^{(\varepsilon)}) = \sum_{w \in W_0} (\tau_w^{(\varepsilon)})^2$.
(iii) $U_\varepsilon^* = U_\varepsilon$.
(iv) $U_\varepsilon = \boldsymbol{c}_+(X^{-\varepsilon}) V_\varepsilon^*$.
(v) *Let* $f, g \in A$. *Then*

$$(U_\varepsilon f, U_\varepsilon g) = \left(\tau_{w_0}^{(\varepsilon)}\right)^{-1} W_0\left(t^{(\varepsilon)}\right)(U_\varepsilon f, g).$$

Proof (i) follows from (5.5.9), by induction on $l(w)$.
(ii) follows from (i).
(iii) By (5.1.22) we have

$$U_\varepsilon^* = \tau_{w_0}^{(\varepsilon)} \sum_{w \in W_0} \left(\tau_w^{(\varepsilon)}\right)^{-1} T(w)^{-1}$$

and since $T(w_0) = T(w_0 w^{-1}) T(w)$, we have

$$T(w)^{-1} = T(w_0)^{-1} T(w_0 w^{-1})$$

giving

$$\begin{aligned} U_\varepsilon^* &= T(w_0)^{-1} \sum_{w \in W_0} \tau_{w_0 w^{-1}}^{(\varepsilon)} T(w_0 w^{-1}) \\ &= \tau_{w_0}^{(\varepsilon)} T(w_0)^{-1} U_\varepsilon = U_\varepsilon \end{aligned}$$

by (i) above.
(iv) follows from (5.5.14) and (iii), since $\boldsymbol{c}_+(X^{-\varepsilon})$ is self-adjoint.
(v) follows from (ii) and (iii). □

5.6 Intertwiners

(5.6.1) *Let $i \in I$. Then $T_i - \boldsymbol{b}_i(X)$ is self-adjoint.*

Proof Since $\boldsymbol{b}_i = \tau_i - \boldsymbol{c}_i$ (4.2.2) we have

$$T_i - \boldsymbol{b}_i(X) = T_i - \tau_i + \boldsymbol{c}_i(X)$$

and by (5.1.22) both $T_i - \tau_i$ and $\boldsymbol{c}_i(X)$ are self-adjoint (since $\boldsymbol{c}_i^* = \boldsymbol{c}_i$). □

Dually, if $i \in I_0$,

$$T_i - \boldsymbol{b}'_i(X)$$

as operator on A', is self-adjoint for the scalar product (5.1.17′).

By (4.3.14),

$$T_i - \boldsymbol{b}'_i(X) = \boldsymbol{c}'_i(X)s_i = s_i\boldsymbol{c}'_i(X^{-1})$$

(where $\boldsymbol{c}'_i = \tau_i - \boldsymbol{b}'_i$), so that

$$\begin{aligned} s_i &= \boldsymbol{c}'_i(X)^{-1}(T_i - \boldsymbol{b}'_i(X)) \\ &= (T_i - \boldsymbol{b}'_i(X))\,\boldsymbol{c}'_i(X^{-1})^{-1} \end{aligned}$$

as operators on A', and hence the adjoint of s_i for this scalar product is

$$\begin{aligned} s_i^{*\prime} &= (T_i - \boldsymbol{b}'_i(X))\,\boldsymbol{c}'_i(X)^{-1} \\ &= \boldsymbol{c}'_i(X^{-1})^{-1}(T_i - \boldsymbol{b}'_i(X)). \end{aligned}$$

As in §5.5, let ε be a linear character of W_0, and define

$$s_i^{(\varepsilon)\prime} = \begin{cases} s_i & \text{if } \varepsilon(s_i) = 1, \\ s_i^{*\prime} & \text{if } \varepsilon(s_i) = -1. \end{cases}$$

Then we have

(5.6.2)
$$\begin{aligned} s_i^{(\varepsilon)\prime} &= (T_i - \boldsymbol{b}'_i(X))\,\boldsymbol{c}'_i\big(X^{-\varepsilon(s_i)}\big)^{-1}, \\ &= \boldsymbol{c}'_i\big(X^{\varepsilon(s_i)}\big)^{-1}(T_i - \boldsymbol{b}'_i(X)). \end{aligned}$$

Let $\eta_i^{(\varepsilon)} = \omega(s_i^{(\varepsilon)\prime})$, where $\omega: \tilde{\mathfrak{H}}' \to \tilde{\mathfrak{H}}$ is the anti-isomorphism defined in (4.7.6). From (5.6.2) we have

$$\begin{aligned}\eta_i^{(\varepsilon)} &= \boldsymbol{c}_i'\left(Y^{\varepsilon(s_i)}\right)^{-1}(T_i - \boldsymbol{b}_i'(Y^{-1})) \\ &= (T_i - \boldsymbol{b}_i'(Y^{-1}))\, \boldsymbol{c}_i'\left(Y^{-\varepsilon(s_i)}\right)^{-1}\end{aligned} \tag{5.6.3}$$

for $i \in I_0$.

It follows from (5.5.2) that if $w \in W_0$ and $w = s_{i_1} \cdots s_{i_p}$ is a reduced expression, then $w^{(\varepsilon)\prime} = s_{i_1}^{(\varepsilon)\prime} \cdots s_{i_p}^{(\varepsilon)\prime}$ and

$$\eta_w^{(\varepsilon)} = \eta_{i_1}^{(\varepsilon)} \cdots \eta_{i_p}^{(\varepsilon)} \tag{5.6.4}$$

are independent of the reduced expression chosen; and from (5.5.3) that

$$\eta_w^{(\varepsilon)} Y^{\lambda'} = Y^{w\lambda'} \eta_w^{(\varepsilon)} \tag{5.6.5}$$

for $w \in W_0$ and $\lambda' \in L'$.

The $\eta_w^{(\varepsilon)}$ are the *Y-intertwiners*. Whereas the elements $w^{(\varepsilon)}$ act as linear operators on all of A, the same is not true of the $\eta_w^{(\varepsilon)}$; since by (4.5.4) $\boldsymbol{c}_i'(r_{k'}(\lambda)) = 0$ if $\lambda = s_i\lambda$, it follows that $\eta_i^{(\varepsilon)}$ acts only on the subspace of A spanned by the E_λ such that $s_i\lambda \neq \lambda$.

(5.6.6) *Let $\lambda \in L, i \in I_0$ and suppose that $<\lambda, \alpha_i^\vee> = r \neq 0$. Then*

$$\eta_i^{(\varepsilon)} E_\lambda = \begin{cases} \tau_i^{-1}\boldsymbol{c}_i'(\varepsilon(s_i)r_{k'}(\lambda))^{-1}E_{s_i\lambda} & \text{if } r > 0, \\ \tau_i\boldsymbol{c}_i'(-\varepsilon(s_i)r_{k'}(\lambda))E_{s_i\lambda} & \text{if } r < 0. \end{cases}$$

Proof Suppose first that $r > 0$. Then

$$\begin{aligned}\eta_i^{(\varepsilon)} E_\lambda &= (T_i - \boldsymbol{b}_i'(Y^{-1}))\, \boldsymbol{c}_i'\left(Y^{-\varepsilon(s_i)}\right)^{-1} E_\lambda \\ &= \boldsymbol{c}_i'(\varepsilon(s_i)r_{k'}(\lambda))^{-1}(T_i - \boldsymbol{b}_i'(r_{k'})\lambda)))E_\lambda \\ &= \tau_i^{-1}\boldsymbol{c}_i'(\varepsilon(s_i)r_{k'}(\lambda))^{-1}E_{s_i\lambda}\end{aligned}$$

by (5.2.2) and (5.4.3).

If now $r < 0$ then $<s_i\lambda, \alpha_i^\vee> > 0$ and hence from above

$$\eta_i^{(\varepsilon)} E_{s_i\lambda} = \tau_i^{-1}\boldsymbol{c}_i'(\varepsilon(s_i)r_{k'}(s_i\lambda))^{-1}E_\lambda.$$

Since $r_{k'}(s_i\lambda) = s_i(r_{k'}(\lambda))$ by (2.8.4), it follows that

$$\eta_i^{(\varepsilon)} E_\lambda = \tau_i\boldsymbol{c}_i'(-\varepsilon(s_i)r_{k'}(\lambda))E_{s_i\lambda}. \qquad \square$$

(5.6.7) *Let $\lambda \in L$. Then*

$$\eta^{(\varepsilon)}_{v(\lambda)} E_\lambda = \left(\xi^{(\varepsilon)}_\lambda\right)^{-1} E_{\lambda_-}$$

where

$$\xi^{(\varepsilon)}_\lambda = \tau_{v(\lambda)} \boldsymbol{c}_{S',\varepsilon k'}(v(\lambda))(r_{k'}(\lambda)).$$

Proof Let $v(\lambda) = s_{i_1} \cdots s_{i_p}$ be a reduced expression, so that

$$\eta^{(\varepsilon)}_{v(\lambda)} = \eta^{(\varepsilon)}_{i_1} \cdots \eta^{(\varepsilon)}_{i_p}.$$

Let

$$\beta_r = s_{i_p} \cdots s_{i_{r+1}}(\alpha_{i_r}), \qquad \lambda_r = s_{i_{r+1}} \cdots s_{i_p}(\lambda)$$

for $0 \le r \le p$, so that $S'_1(v(\lambda)) = \{\beta_1^\vee, \dots, \beta_p^\vee\}$ and

$$<\lambda_r, \alpha_{i_r}^\vee> = <\lambda, \beta_r^\vee> > 0$$

by (2.4.4). Hence by (5.6.6)

$$\eta^{(\varepsilon)}_{v(\lambda)} E_\lambda = \xi^{-1} E_{\lambda_-}$$

where

$$\xi = \prod_{r=1}^{p} \tau_{i_r} \boldsymbol{c}'_{i_r}\left(\varepsilon\left(s_{i_r}\right) r_{k'}(\lambda_r)\right),$$

and since $r_{k'}(\lambda_r) = s_{i_{r+1}} \cdots s_{i_p} r_{k'}(\lambda)$ by (2.8.4), we have

$$\begin{aligned} \boldsymbol{c}'_{i_r}\left(\varepsilon\left(s_{i_r}\right) r_{k'}(\lambda_r)\right) &= \boldsymbol{c}\left(\tau_{i_r}, \upsilon_{i_r}; q^{\varepsilon(s_{i_r})<r_{k'}(\lambda_r), \alpha_{i_r}^\vee>}\right) \\ &= \boldsymbol{c}\left(\tau_{i_r}^{\varepsilon(s_{i_r})}, \upsilon_{i_r}^{\varepsilon(s_{i_r})}; q^{<r_{k'}(\lambda), \beta_r^\vee>}\right). \\ &= \boldsymbol{c}_{\beta_r^\vee, \varepsilon k'}(r_{k'}(\lambda)). \end{aligned}$$

Hence

$$\xi = \tau_{v(\lambda)} \boldsymbol{c}_{S',\varepsilon k'}(v(\lambda))(r_{k'}(\lambda)).$$

□

Finally, let

(5.6.8)
$$\begin{aligned} V'_\varepsilon &= \varepsilon(w_0) \sum_{w \in W_0} \varepsilon(w) w^{(\varepsilon)\prime}, \\ \mathscr{V}_\varepsilon &= \omega(V'_\varepsilon) = \varepsilon(w_0) \sum_{w \in W_0} \varepsilon(w) \eta^{(\varepsilon)}_w. \end{aligned}$$

As in (5.5.14) we have

(5.6.9) $$U_\varepsilon = V'_\varepsilon c'_+(X^{-\varepsilon}),$$

where

(5.6.10) $$c'_+(X^{-\varepsilon}) = \prod_{\alpha \in R^+} c_{\alpha^\vee,k'}\left(X^{-\varepsilon(s_\alpha)}\right).$$

Applying ω to (5.6.9) gives

(5.6.11) $$U_\varepsilon = c'_+(Y^\varepsilon)\mathscr{V}_\varepsilon = \mathscr{V}^*_\varepsilon c'_+(Y^\varepsilon)$$

since $\omega(U_\varepsilon) = U_\varepsilon$ and $U^*_\varepsilon = U_\varepsilon$ by (5.5.17). From (5.6.3) we have

$$\left(\eta_i^{(\varepsilon)}\right)^* = \left(T_i - b'_i(Y^{-1})\right)c'_i\left(Y^{\varepsilon(s_i)}\right)^{-1}$$

since both $T_i - b'_i(Y^{-1})$ and $c'_i(Y^{\varepsilon(s_i)})$ are self adjoint. Thus

(5.6.12) $$\left(\eta_i^{(\varepsilon)}\right)^* = \eta_i^{(-\varepsilon)}$$

where $-\varepsilon$ is the character $w \mapsto (-1)^{l(w)}\varepsilon(w)$ of W_0. Hence

(5.6.13) $$\mathscr{V}^*_\varepsilon = \varepsilon(w_0)\sum_{w \in W_0} \varepsilon(w)\eta_w^{(-\varepsilon)}.$$

The operators $c'_+(Y^\varepsilon)\eta_i^{(\varepsilon)}$ are well-defined as operators on A. We have

(5.6.14) $$c'_+(Y^\varepsilon)\eta_i^{(\varepsilon)}E_\lambda = \eta_i^{(-\varepsilon)}c'_+(Y^\varepsilon)E_\lambda = 0$$

if $\lambda = s_i\lambda$. For $c'_+(Y^\varepsilon)\eta_i^{(\varepsilon)}E_\lambda$ is of the form $f(Y)(T_i - b'_i(Y^{-1}))E_\lambda$ which is zero by (5.4.2).

5.7 The polynomials $P_\lambda^{(\varepsilon)}$

As before, let ε be a linear character of W_0. For each $\lambda \in L$ we define

$$F_\lambda^{(\varepsilon)} = U_\varepsilon E_\lambda.$$

(5.7.1) *Let $i \in I_0$. If $\varepsilon(s_i) = -1$ and $\lambda = s_i\lambda$, then $F_\lambda^{(\varepsilon)} = 0$.*

Proof By (5.4.4) and (5.5.9) we have

$$\tau_i F_\lambda^{(\varepsilon)} = \tau_i U_\varepsilon E_\lambda = U_\varepsilon T_i E_\lambda = T_i U_\varepsilon E_\lambda = -\tau_i^{-1} F_\lambda^{(\varepsilon)}.$$ □

(5.7.2) *If* $<\lambda, \alpha_i'> > 0$ *then*

$$F_{s_i\lambda}^{(\varepsilon)} = \varepsilon(s_i)\tau_i \boldsymbol{c}_i'(\varepsilon(s_i)r_{k'}(\lambda))F_\lambda^{(\varepsilon)}.$$

Proof By (5.4.3) and (5.5.9) we have

$$\begin{aligned}\tau_i^{(\varepsilon)} F_\lambda^{(\varepsilon)} &= U_\varepsilon T_i E_\lambda \\ &= U_\varepsilon(\tau_i^{-1} E_{s_i\lambda} + \boldsymbol{b}_i'(r_{k'}(\lambda))E_\lambda) \\ &= \tau_i^{-1} F_{s_i\lambda}^{(\varepsilon)} + \boldsymbol{b}_i'(r_{k'}(\lambda))F_\lambda^{(\varepsilon)},\end{aligned}$$

so that

$$\begin{aligned}F_{s_i\lambda}^{(\varepsilon)} &= \tau_i\left(\tau_i^{(\varepsilon)} - \boldsymbol{b}_i'(r_{k'}(\lambda))\right)F_\lambda^{(\varepsilon)} \\ &= \varepsilon(s_i)\tau_i \boldsymbol{c}_i'(\varepsilon(s_i)r_{k'}(\lambda))F_\lambda^{(\varepsilon)}.\end{aligned}$$

by (4.2.3). □

In view of (5.7.2), we may assume that $\lambda \in L$ is dominant, since $F_\mu^{(\varepsilon)}$ for $\mu \in W_0\lambda$ is a scalar multiple of $F_\lambda^{(\varepsilon)}$ and hence $U_\varepsilon A_\lambda$ has dimension at most 1. Also, in view of (5.7.1), we shall assume henceforth that

(5.7.3) $\varepsilon(w) = 1$ *for all* $w \in W_{0\lambda}$,

where $W_{0\lambda}$ is the subgroup of W_0 that fixes λ.

When ε is the trivial character, this is no restriction. On the other hand, when ε is the sign character, (5.7.3) requires that λ is *regular* dominant.

(5.7.4) (i) $F_\lambda^{(\varepsilon)}$ *is* $W_{0\lambda}$*-symmetric.*
(ii) *When* ε *is the trivial character,* $F_\lambda^{(\varepsilon)}$ *is* W_0*-symmetric.*

Proof If $\varepsilon(s_i) = 1$ we have $(T_i - \tau_i)F_\lambda^{(\varepsilon)} = 0$ by (5.5.9), and hence $s_i F_\lambda^{(\varepsilon)} = F_\lambda^{(\varepsilon)}$ by (4.3.12). □

Each coset $wW_{0\lambda}$ has a unique element of minimal length, namely $\bar{v}(\mu)$ in the notation of §2.7, where $\mu = w\lambda$. Let

$$W_0^\lambda = \{\bar{v}(\mu) : \mu \in W_0\lambda\}.$$

Then every element of W_0 is uniquely of the form vw, where $v \in W_0^\lambda$ and $w \in W_{0\lambda}$, and $l(vw) = l(v) + l(w)$. Hence

$$U_\varepsilon = \left(\tau_{w_0}^{(\varepsilon)}\right)^{-1} \left(\sum_{v \in W_0^\lambda} \tau_v^{(\varepsilon)} T(v) \right) \left(\sum_{w \in W_{0\lambda}} \tau_w T(w) \right),$$

and since $T(w)E_\lambda = \tau_w E_\lambda$ for $w \in W_{0\lambda}$, it follows that

(5.7.5) $$F_\lambda^{(\varepsilon)} = \left(\tau_{w_0}^{(\varepsilon)}\right)^{-1} W_{0\lambda}(\tau^2) \sum_{v \in W_0^\lambda} \tau_v^{(\varepsilon)} T(v) E_\lambda,$$

where

(5.7.6) $$W_{0\lambda}(\tau^2) = \sum_{w \in W_{0\lambda}} \tau_w^2.$$

The only term on the right-hand side of (5.7.5) that contains $e^{w_0\lambda}$ is that corresponding to $v = v(\lambda)$, the shortest element of W_0 that takes λ to $w_0\lambda$. By (5.4.3) the coefficient of $e^{w_0\lambda}$ in $T(v(\lambda))E_\lambda$ is $\tau_{v(\lambda)}^{-1}$, and hence the coefficient of $e^{w_0\lambda}$ in $F_\lambda^{(\varepsilon)}$ is $\tau_{w_0}^{-1} W_{0\lambda}(\tau^2)$, since by (5.7.3) $\tau_{v(\lambda)}^{(\varepsilon)}/\tau_{v(\lambda)} = \tau_{w_0}^{(\varepsilon)}/\tau_{w_0}$.

Accordingly we define (always for $\lambda \in L$ dominant)

(5.7.7) $$\begin{aligned} P_\lambda^{(\varepsilon)} &= \tau_{w_0} W_{0\lambda}(\tau^2)^{-1} F_\lambda^{(\varepsilon)} \\ &= e^{w_0\lambda} + \text{lower terms.} \end{aligned}$$

In terms of the E_μ we have

(5.7.8) $$P_\lambda^{(\varepsilon)} = \sum_{\mu \in W_0\lambda} \varepsilon(v(\mu))\, \xi_\mu^{(-\varepsilon)} E_\mu$$

where $\xi_\mu^{(-\varepsilon)}$ *is given by* (5.6.7) (*with* $-\varepsilon$ *replacing* ε).

Proof From (5.7.2) it follows that $P_\lambda^{(\varepsilon)}$ is proportional to

$$U_\varepsilon E_{w_0\lambda} = \varepsilon(w_0) \sum_{w \in W_0} \varepsilon(w) \eta_w^{(-\varepsilon)} \boldsymbol{c}'_+(Y^\varepsilon) E_{w_0\lambda}$$

by (5.6.11) and (5.6.13). By (5.6.14), only the elements of W_0 of the form $v(\mu)^{-1}$, where $\mu \in W_0\lambda$, contribute to this sum. Since $\boldsymbol{c}'_+(Y^\varepsilon)E_{w_0\lambda}$ is a scalar multiple of $E_{w_0\lambda}$, it follows that $P_\lambda^{(\varepsilon)}$ is proportional to

$$\sum_{\mu \in W_0\lambda} \varepsilon(v(\mu)) \left(\eta_{v(\mu)}^{(-\varepsilon)}\right)^{-1} E_{w_0\lambda}$$

which by (5.6.7) is equal to

$$\sum_{\mu \in W_0\lambda} \varepsilon(v(\mu)) \xi_\mu^{(-\varepsilon)} E_\mu.$$

Since the coefficient of $E_{w_0\lambda}$ in this sum is equal to 1 (because $v(\mu) = 1$ when $\mu = w_0\lambda$), (5.7.8) is proved. □

(5.7.9) *Let $f \in A'_0$. Then*

$$f(Y)P_\lambda^{(\varepsilon)} = f(-\lambda - \rho_{k'})P_\lambda^{(\varepsilon)}.$$

Proof Since $f(Y)$ commutes with $T(w)$ for each $w \in W_0$ by (4.2.10), it commutes with U_ε. Hence

$$\begin{aligned} f(Y)U_\varepsilon E_{w_0\lambda} &= U_\varepsilon f(Y)E_{w_0\lambda} \\ &= f(-\lambda - \rho_{k'})\, U_\varepsilon E_{w_0\lambda} \end{aligned}$$

by (5.2.2). Since $U_\varepsilon E_{w_0\lambda}$ is proportional to $P_\lambda^{(\varepsilon)}$, the result follows. □

From (5.7.9) and (5.7.4) it follows that when ε is the trivial character of W_0,

(5.7.10) $$P_\lambda^{(\varepsilon)} = P_\lambda$$

as defined in §5.3. Also from (5.7.9) it follows, exactly as in §5.2 and §5.3, that the $P_\lambda^{(\varepsilon)}$ are pairwise orthogonal:

(5.7.11) $$\left(P_\lambda^{(\varepsilon)}, P_\mu^{(\varepsilon)}\right) = 0$$

if $\lambda \neq \mu$.

(5.7.12) *Let $\lambda \in L_{++}$. Then*

$$\left(P_\lambda^{(\varepsilon)}, P_\lambda^{(\varepsilon)}\right)/(P_\lambda, P_\lambda) = \xi_\lambda^{(-\varepsilon)}/\xi_\lambda^{(-1)}$$

where -1 denotes the sign character of W_0.

Proof From (5.7.7) and (5.5.17) (v) we have

$$\begin{aligned} \left(P_\lambda^{(\varepsilon)}, P_\lambda^{(\varepsilon)}\right) &= \frac{(U_\varepsilon E_\lambda, U_\varepsilon E_\lambda)}{W_{0\lambda}(\tau^2)W_{0\lambda}(\tau^{-2})} \\ &= \frac{W_0\left(\left(\tau^{(\varepsilon)}\right)^2\right)(U_\varepsilon E_\lambda, E_\lambda)}{\tau_{w_0}^{(\varepsilon)} W_{0\lambda}(\tau^2)W_{0\lambda}(\tau^{-2})} \\ &= \frac{W_0\left(\left(\tau^{(\varepsilon)}\right)^2\right)\left(P_\lambda^{(\varepsilon)}, E_\lambda\right)}{\tau_{w_0}\tau_{w_0}^{(\varepsilon)} W_{0\lambda}(\tau^{-2})} \\ &= \frac{\varepsilon(v(\lambda))W_0\left(\left(\tau^{(\varepsilon)}\right)^2\right)\xi_\lambda^{(-\varepsilon)}(E_\lambda, E_\lambda)}{\tau_{w_0}\tau_{w_0}^{(\varepsilon)} W_{0\lambda}(\tau^{-2})} \end{aligned}$$

by (5.7.8) and orthogonality of the E_μ. Hence

$$\frac{(P_\lambda^{(\varepsilon)}, P_\lambda^{(\varepsilon)})}{(P_\lambda, P_\lambda)} = \frac{\varepsilon(v(\lambda)) W_0\big((\tau^{(\varepsilon)})^2\big)\tau_{w_0}}{W_0(\tau^2)\tau_{w_0}^{(\varepsilon)}} \cdot \frac{\xi_\lambda^{(-\varepsilon)}}{\xi_\lambda^{(-1)}}.$$

Let

$$R_1 = \{\alpha \in R : \varepsilon(s_\alpha) = +1\},$$
$$R_{-1} = \{\alpha \in R : \varepsilon(s_\alpha) = -1\},$$

and let W_1 (resp. W_{-1}) be the subgroup of W_0 generated by the s_i such that $\varepsilon(s_i) = +1$ (resp. -1). Then the kernel of ε is the Weyl group of R_1, and is the normal closure $\bar{W}_1$ of W_1 in W_0; and W_0 is the semidirect product

$$W_0 = \bar{W}_1 \rtimes W_{-1}.$$

Hence

(1) $$W_0\big((\tau^{(\varepsilon)})^2\big) = \bar{W}_1(\tau^2)W_{-1}(\tau^{-2})$$

in an obvious notation; also $w_0 = w_1 w_{-1}$ where w_1 (resp. w_{-1}) is the longest element of $\bar{W}_1$, (resp. W_{-1}) so that

(2) $$\tau_{w_0}^{(\varepsilon)} = \varepsilon(w_0)\tau_{w_1}\tau_{w_{-1}}^{-1}.$$

Finally, $v(\lambda) = w_0 w_{0\lambda}$, where $w_{0\lambda}$ is the longest element of $W_{0\lambda}$. Since $\varepsilon(w) = 1$ for $w \in W_{0\lambda}$ (5.7.3), it follows that

(3) $$\varepsilon(v(\lambda)) = \varepsilon(w_0).$$

From (1), (2) and (3) it follows that

$$\frac{\varepsilon(v(\lambda)) W_0\big((\tau^{(\varepsilon)})^2\big)\tau_{w_0}}{W_0(\tau^2)\tau_{w_0}^{(\varepsilon)}} = 1,$$

completing the proof of (5.7.12). □

Suppose in particular that ε is the sign character of W_0. In that case we write

(5.7.13) $$Q_\lambda = P_\lambda^{(\varepsilon)}$$

for $\lambda \in L$ regular dominant. Then we obtain from (5.7.12) and the definition (5.6.7) of $\xi_\lambda^{(\varepsilon)}$

(5.7.14) $$\frac{(Q_\lambda, Q_\lambda)}{(P_\lambda, P_\lambda)} = \prod_{\alpha \in R^+} \frac{c_{\alpha^\vee, k'}(\lambda + \rho_{k'})}{c_{\alpha^\vee, -k'}(\lambda + \rho_{k'})}$$

since $v(\lambda) = w_0$ and $r_{k'}(\lambda) = \lambda + \rho_{k'}$, by (2.8.2′).

5.8 Norms

As before, S is an irreducible affine root system as in (1.4.1)–(1.4.3). Recall that

$$S_0 = \{a \in S : a(0) = 0\}, \qquad S_1 = \{a \in S : \tfrac{1}{2}a \notin S\}.$$

Let ε be a linear character of W_0, and let l be the labelling of S such that $l(a) = 1$ if s_a is conjugate in W to s_i, where $i \neq 0$ and $\varepsilon(s_i) = -1$; and $l(a) = 0$ otherwise. Let $k + l$ denote the labelling $a \mapsto k(a) + l(a)$, and as before let $(\varepsilon k)(a) = \varepsilon(s_a)k(a)$ for $a \in S_0$.

For each $a \in S_1$, let

$$(5.8.1) \qquad \begin{aligned} \delta_a = \delta_{a,k} &= q^{k(a)/2}e^{a/2} - q^{-k(a)/2}e^{-a/2} \\ &= \left(e^{a/2} - e^{-a/2}\right)\boldsymbol{c}_{a,k} \end{aligned}$$

if $2a \notin S$, and

$$(5.8.2) \qquad \begin{aligned} \delta_a = \delta_{a,k} &= \left(q^{k(a)/2}e^{a/2} - q^{-k(a)/2}e^{-a/2}\right) \\ &\quad \times \left(q^{k(2a)/2}e^{a/2} + q^{-k(2a)/2}e^{-a/2}\right) \\ &= (e^{a} - e^{-a})\boldsymbol{c}_{a,k} \end{aligned}$$

if $2a \in S$.

Let

$$(5.8.3) \qquad \delta_{\varepsilon,k} = \prod_{\substack{a \in S_{01}^+ \\ l(a)=1}} \delta_{a,k}$$

where $S_{01}^+ = S_0 \cap S_1 \cap S^+$. Then we have

$$(5.8.4) \qquad \delta_{\varepsilon,k}\delta_{\varepsilon,k}^*\Delta_{S,k} = \nabla_{S,k+l}/\Delta_{S,\varepsilon k}^0.$$

Proof Suppose that $S = S(R)$ as in (1.4.1). Then

$$\begin{aligned} \Delta_{S,k+l}/\Delta_{S,k} &= \prod_{\substack{\alpha \in R^+ \\ l(\alpha)=1}} \left(1 - q^{k(\alpha)}e^{\alpha}\right)\left(1 - q^{k(\alpha)+1}e^{-\alpha}\right) \\ &= \delta_{\varepsilon,k}\delta_{\varepsilon,k}^* \prod_{\substack{\alpha \in R^+ \\ l(\alpha)=1}} \frac{1 - q^{k(\alpha)+1}e^{-\alpha}}{1 - q^{-k(\alpha)}e^{-\alpha}} \\ &= \delta_{\varepsilon,k}\delta_{\varepsilon,k}^* \prod_{\alpha \in R^+} \frac{1 - q^{k(\alpha)+l(\alpha)}e^{-\alpha}}{1 - q^{(\varepsilon k)(\alpha)}e^{-\alpha}} \\ &= \delta_{\varepsilon,k}\delta_{\varepsilon,k}^*\Delta_{\varepsilon k}^0/\Delta_{k+l}^0. \end{aligned}$$

Likewise in the other two cases (1.4.2), (1.4.3). □

If k is any labelling of S and τ_i $(i\in I)$ are defined as in (5.1.6), and τ_w $(w\in W_0)$ as in (4.5.4), we shall write

(5.8.5) $$W_0(q^k) = \sum_{w\in W_0} \tau_w^2.$$

We shall also denote the scalar product (5.1.17) by $(f, g)_k$:

$$(f, g)_k = \mathrm{ct}(fg^*\Delta_{S,k})$$

since from now on several labellings will be in play.

(5.8.6) *Let $f, g \in A_0$. Then*

$$(f, g)_{k+l} = \frac{W_0(q^{k+l})}{W_0(q^{\varepsilon k})}(\delta_{\varepsilon,k} f, \delta_{\varepsilon,k} g)_k.$$

Proof We have

$$\begin{aligned}(\delta_{\varepsilon,k} f, \delta_{\varepsilon,k} g)_k &= \mathrm{ct}(fg^*\delta_{\varepsilon,k}\delta^*_{\varepsilon,k}\Delta_{S,k})\\ &= \mathrm{ct}\big(fg^*\nabla_{S,k+l}\big(\Delta^0_{\varepsilon,k}\big)^{-1}\big)\\ &= W_0(q^{\varepsilon k}) <f, g^0>_{k+l}\\ &= \frac{W_0(q^{\varepsilon k})}{W_0(q^{k+l})}(f, g)_{k+l}\end{aligned}$$

by (5.8.4) and (5.1.34). □

(5.8.7) *For each $i \in I_0$ we have*
(i) $(T_i - \tau_i^{(\varepsilon)})\delta_{\varepsilon,k}(X) = (s_i\delta_{\varepsilon,k})(X)(T_i - \tau_i)$,
(ii) $(T_i - \tau_i)\delta_{\varepsilon,k}(X^{-1}) = (s_i\delta_{\varepsilon,k})(X^{-1})(T_i - \tau_i^{(\varepsilon)})$.

Proof (i) By (4.3.15) we have

$$T_i\delta_{\varepsilon,k}(X) - (s_i\delta_{\varepsilon,k})(X)T_i = \boldsymbol{b}_i(X)(\delta_{\varepsilon,k}(X) - (s_i\delta_{\varepsilon,k})(X)).$$

If $\varepsilon(s_i) = 1$ this is zero, since s_i permutes the $a \in S_{01}^+$ such that $l(a) = 1$ and hence fixes $\delta_{\varepsilon,k}$. If on the other hand $\varepsilon(s_i) = -1$, then by (4.2.3)

$$\begin{aligned}&T_i\delta_{\varepsilon,k}(X) - (s_i\delta_{\varepsilon,k})(X)T_i\\ &\quad= \big(\boldsymbol{c}_i(X^{-1}) - \tau_i^{-1}\big)\delta_{\varepsilon,k}(X) + \big(\boldsymbol{c}_i(X) - \tau_i\big)(s_i\delta_{\varepsilon,k})(X)\\ &\quad= -\tau_i^{-1}\delta_{\varepsilon,k}(X) - \tau_i(s_i\delta_{\varepsilon,k})(X)\end{aligned}$$

because $\delta_{\varepsilon,k}/s_i\delta_{\varepsilon,k} = \delta_{a_i,k}/\delta_{-a_i,k} = -\boldsymbol{c}_i/\bar{\boldsymbol{c}}_i$.

The proof of (ii) is similar. □

Next we have

(5.8.8) $U_\varepsilon A = \delta_{\varepsilon,k} A_0$.

Proof Let $f \in U_\varepsilon A$. By (5.5.9), $(T_i - \tau_i^{(\varepsilon)})f = 0$ for all $i \neq 0$. Hence (5.8.7) (i) shows that $g = \delta_{\varepsilon,k}^{-1} f$ is killed by $T_i - \tau_i$ for each $i \neq 0$, and hence is W_0-symmetric. Consequently $w_0(\delta_{\varepsilon,k}^{-1} f) = \delta_{\varepsilon,k}^{-1} f$, i.e.,

$$\delta_{\varepsilon,k} w_0(f) = w_0(\delta_{\varepsilon,k}) f.$$

Now $\delta_{\varepsilon,k}$ and $w_0(\delta_{\varepsilon,k})$ are coprime elements of A. Hence $\delta_{\varepsilon,k}$ divides f in A, so that $g \in A_0$ and $f \in \delta_{\varepsilon,k} A_0$.

Conversely, if $f \in \delta_{\varepsilon,k} A_0$, then $(T_i - \tau_i^{(\varepsilon)})f = 0$ for all $i \neq 0$ by (5.8.7) (i), and therefore $f \in U_\varepsilon A$ by (5.5.10) (ii). □

(5.8.9) *Let $\lambda \in L_{++}$. Then*

$$P_{\lambda+\rho_l,k}^{(\varepsilon)} = \varepsilon(w_0) q^{n(k,l)/2} \delta_{\varepsilon,k} P_{\lambda,k+l},$$

where

$$n(k,l) = \frac{1}{2} \sum_{a \in S_0^+} k(a) l(a)$$

and

$$\rho_l = \frac{1}{2} \sum_{a \in S_{01}^+} l(a) u_a a$$

where $u_a = 1$ if $2a \notin S$, and $u_a = 2$ if $2a \in S$.

Proof Since $P_{\lambda+\rho_l,k}^{(\varepsilon)} \in U_\varepsilon A$, it follows from (5.8.8) that $P_{\lambda+\rho_l,k}^{(\varepsilon)} = \delta_{\varepsilon,k} g$ for some $g \in A_0$. The leading term in $P_{\lambda+\rho_l,k}^{(\varepsilon)}$ is $e^{w_0\lambda - \rho_l}$, and in $\delta_{\varepsilon,k}$ is $\varepsilon(w_0) q^{-n(k,l)/2} e^{-\rho_l}$. Hence

(1) $$g = \varepsilon(w_0) q^{n(k,l)/2} m_\lambda + \text{ lower terms.}$$

Let $\mu \in L_{++}$, $\mu < \lambda$. The highest exponential that occurs in $\delta_{\varepsilon,k} m_\mu$ is $e^{w_0(\mu+\rho_l)}$. Since $P_{\lambda+\rho_l,k}^{(\varepsilon)}$ is a linear combination of the $E_{w(\lambda+\rho_l)}$, $w \in W_0$, it follows that

$$(\delta_{\varepsilon,k} g, \delta_{\varepsilon,k} m_\mu)_k = 0$$

and hence by (5.8.6) that

(2) $$(g, m_\mu)_{k+l} = 0$$

for all $\mu \in L_{++}$ such that $\mu < \lambda$. From (1) and (2) we conclude that $g = \varepsilon(w_0)q^{n(k,l)/2}P_{\lambda,k+l}$. □

In particular, when $\lambda = 0$ we have

(5.8.10) $$P^{(\varepsilon)}_{\rho_l,k} = \varepsilon(w_0)q^{n(k,l)/2}\delta_{\varepsilon,k}$$

and therefore, for any $\lambda \in L_{++}$,

(5.8.11) $$P_{\lambda,k+l} = P^{(\varepsilon)}_{\lambda+\rho_l,k}/P^{(\varepsilon)}_{\rho_l,k}.$$

Thus when ε is the sign character of W_0 we have

(5.8.12) $$P_{\lambda,k+1} = Q_{\lambda+\rho,k}/Q_{\rho,k}$$

where $k+1$ is the labelling $a \mapsto k(a)+1$ of S, and

(5.8.13) $$\rho = \frac{1}{2}\sum_{a\in S_{01}^+} u_a a.$$

(In the cases (1.4.1) and (1.4.3), $\rho = \frac{1}{2}\sum_{\alpha\in R^+}\alpha$; in the case (1.4.2), $\rho = \frac{1}{2}\sum_{\alpha\in R^+}\alpha^\vee$.)

Remark (5.8.12) may be regarded as a generalization of Weyl's character formula, which is the case $k = 0$: for then $E_\lambda = e^\lambda$ for all $\lambda \in L$, and $Q_{\lambda+\rho,0} = \varepsilon(w_0)\sum_{w\in W_0}\varepsilon(w)e^{w(\lambda+\rho)}$.

From (5.8.6) and (5.8.9) we have

$$\begin{aligned}\left(P^{(\varepsilon)}_{\lambda+\rho_l,k}, P^{(\varepsilon)}_{\lambda+\rho_l,k}\right)_k &= (\delta_{\varepsilon,k}P_{\lambda,k+l}, \delta_{\varepsilon,k}P_{\lambda,k+l})_k \\ &= \frac{W_0(q^{\varepsilon k})}{W_0(q^{k+l})}(P_{\lambda,k+l}, P_{\lambda,k+l})_{k+l}\end{aligned}$$

and therefore, by (5.7.12),

$$\frac{(P_{\lambda,k+l}, P_{\lambda,k+l})_{k+l}}{(P_{\lambda+\rho_l,k}, P_{\lambda+\rho_l,k})_k} = \frac{W_0(q^{k+l})\xi^{(-\varepsilon)}_{\lambda+\rho_l}}{W_0(q^{\varepsilon k})\xi^{(-1)}_{\lambda+\rho_l}},$$

Equivalently, by (5.1.34) and (5.3.2),

(5.8.14) $$\frac{|P_{\lambda,k+l}|^2_{k+l}}{|P_{\lambda+\rho_l,k}|^2_k} = \frac{W_0(q^k)\xi^{(-\varepsilon)}_{\lambda+\rho_l}}{W_0(q^{\varepsilon k})\xi^{(-1)}_{\lambda+\rho_l}},$$

where

$$|f|_k^2 = \langle f, f \rangle_k$$

for $f \in A$.

The right-hand side of (5.8.14) can be reformulated, as follows. Let $\mu = \lambda + \rho_l$. Then we have, in the notation of (5.3.11),

(5.8.15) $$\frac{W_0(q^k)\xi_\mu^{(-\varepsilon)}}{W_0(q^{\varepsilon k})\xi_\mu^{(-1)}} = \frac{\Delta^+_{S',k'+l'}(\mu+\rho_{k'})\Delta^-_{S',-k'-l'}(-\mu-\rho_{k'})}{\Delta^+_{S',k'}(\mu+\rho_{k'})\Delta^-_{S',-k'}(-\mu-\rho_{k'})}.$$

Proof We shall verify (5.8.15) when $S = S(R)$ (1.4.1); the other cases are analogous. Consider first the right-hand side. From (5.1.12) we have

$$\Delta^+_{S',k'+l'}/\Delta^+_{S',k'} = \prod_{\substack{\alpha\in R^+\\ l(\alpha)=1}} \left(1 - q^{k(\alpha)}e^{\alpha^\vee}\right)$$

and

$$\Delta^-_{S',-k'-l'}/\Delta^-_{S',-k'} = \prod_{\substack{\alpha\in R^+\\ l(\alpha)=1}} \left(1 - q^{-k(\alpha)}e^{-\alpha^\vee}\right)^{-1},$$

so that the right-hand side of (5.8.15) is equal to

(1) $$\prod_{\substack{\alpha\in R^+\\ l(\alpha)=1}} \frac{1 - q^{k(\alpha)+\langle\mu+\rho_k,\alpha^\vee\rangle}}{1 - q^{-k(\alpha)+\langle\mu+\rho_k,\alpha^\vee\rangle}}.$$

Next, consider the left-hand side of (5.8.15). From (2.4.4) we have

$$S'(v(\mu)) = \{\alpha^\vee \in (R^\vee)^+ : \langle\mu,\alpha^\vee\rangle > 0\},$$

so that by (5.6.7)

$$\frac{\xi_\mu^{(-\varepsilon)}}{\xi_\mu^{(-1)}} = \prod_{\substack{\alpha\in R^+\\ \langle\mu,\alpha^\vee\rangle>0}} \frac{\boldsymbol{c}_{\alpha^\vee,-\varepsilon k}(r_k(\mu))}{\boldsymbol{c}_{\alpha^\vee,-k}(r_k(\mu))}.$$

Since μ is dominant,

$$r_k(\mu) = w_{0\mu}(\mu + \rho_k)$$

by (2.8.7), where $w_{0\mu}$ is the longest element of the isotropy group $W_{0\mu}$ of μ in W_0. Since $\langle\mu, w_{0\mu}\alpha^\vee\rangle = \langle\mu,\alpha^\vee\rangle$, it follows that $w_{0\mu}$ permutes the roots

$\alpha^\vee$ such that $<\mu, \alpha^\vee> > 0$. Hence

$$\frac{\xi_\mu^{(-\varepsilon)}}{\xi_\mu^{(-1)}} = \prod_{\substack{\alpha\in R^+ \\ <\mu,\alpha^\vee>>0}} \frac{c_{\alpha^\vee,-\varepsilon k}(\mu+\rho_k)}{c_{\alpha^\vee,-k}(\mu+\rho_k)}.$$

The terms in this product corresponding to roots $\alpha \in R^+$ such that $\varepsilon(s_\alpha) = 1$ (i.e., $l(\alpha) = 0$) are equal to 1. Hence we may assume that $l(\alpha) = 1$, which by (5.7.3) implies that $<\mu, \alpha^\vee> > 0$. Hence finally we have

$$\text{(2)} \qquad \frac{\xi_\mu^{(-\varepsilon)}}{\xi_\mu^{(-1)}} = \prod_{\substack{\alpha\in R^+ \\ l(\alpha)=1}} q^{-k(\alpha)} \frac{1-q^{k(\alpha)+<\mu+\rho_k,\alpha^\vee>}}{1-q^{-k(\alpha)+<\mu+\rho_k,\alpha^\vee>}}.$$

As in §5.7 we have

$$W_0(q^{\varepsilon k}) = \bar{W}_1(q^k)W_{-1}(q^{-k}),$$
$$W_0(q^k) = \bar{W}_1(q^k)W_{-1}(q^k),$$

so that

$$W_0(q^k)/W_0(q^{\varepsilon k}) = W_{-1}(q^k)/W_{-1}(q^{-k})$$
$$\text{(3)} \qquad = \prod_{\substack{\alpha\in R^+ \\ l(\alpha)=1}} q^{k(\alpha)}.$$

From (2) and (3) we see that the left-hand side of (5.8.15) is equal to (1). □

From (5.8.14) and (5.8.15) we have

$$\text{(5.8.16)} \qquad \frac{|P_{\lambda,k+l}|^2_{k+l}}{|P_{\lambda+\rho_l,k}|^2_k} = \frac{\Delta^+_{S',k'+l'}(\lambda+\rho_{k'+l'})\Delta^-_{S',-k'-l'}(-\lambda-\rho_{k'+l'})}{\Delta^+_{S',k'}(\lambda+\rho_{k'+l'})\Delta^-_{S',-k'}(-\lambda-\rho_{k'+l'})}.$$

This provides the inductive step in the proof of the *norm formula*:

$$\text{(5.8.17)} \qquad |P_{\lambda,k}|^2_k = \Delta^+_{S',k'}(\lambda+\rho_{k'})\Delta^-_{S',-k'}(-\lambda-\rho_{k'}).$$

Proof (a) Suppose first that $S = S(R)$ (1.4.1). If $k(\alpha) = 1$ for all $\alpha \in R$, then $P_{\lambda,k} = \chi_{R,\lambda}$ and $|P_{\lambda,k}|^2_k = 1$ for all $\lambda \in L_{++}$ by (5.3.15). On the other hand, it follows from the definitions that

$$\Delta^+_{S',k'}\Delta^-_{S',-k'} = \prod_{\alpha\in R^+}(1-e^{\alpha^\vee})/(1-e^{-\alpha^\vee})$$

so that $\Delta^+_{S',k'}(\lambda+\rho_{k'})\Delta^-_{S',-k'}(-\lambda-\rho_{k'}) = 1$.

Hence (5.8.17) is true when all the labels $k(\alpha)$ are equal to 1. But now (5.8.16) shows that the norm formula is true for (λ, k) if it is true for $(\lambda+\rho_l, k-l)$. Hence it is true whenever the labels $k(\alpha)$ are positive integers.

(b) Suppose next that $S = S(R)^\vee$ (1.4.2). If $k^\vee(\alpha) = 1$ for all $\alpha \in R$, in the notation of (5.1.13), then again $P_{\lambda,k} = \chi_{R^\vee,\lambda}$ and $|P_{\lambda,k}|_k^2 = 1$ for all $\lambda \in L_{++}$ (5.3.15), and the conclusion is the same as before: (5.8.17) is true whenever the $k^\vee(\alpha)$ are positive integers.

(c) Finally, suppose that S is of type $(C_n^\vee, C_n)$ (1.4.3), and that the labels $k(a)$ are positive integers. By (5.1.28) (iii), $\nabla_{S,k}$ (and therefore also $P_{\lambda,k}$) is symmetrical in $u_1, \ldots, u_4$, where

$$(u_1, u_2, u_3, u_4) = \left(q^{k_1}, -q^{k_2}, q^{\frac{1}{2}+k_3}, -q^{\frac{1}{2}+k_4}\right).$$

Let $l = (1, 1, 0, 0)$. Then (5.8.16) shows that the norm formula is true for the parameters (u_1, u_2, u_3, u_4) if and only if it is true for the parameters $(q^{-1}u_1, q^{-1}u_2, u_3, u_4)$. Hence by symmetry it is true for (u_1, u_2, u_3, u_4) if and only if it is true when any two of the u_i are replaced by $q^{-1}u_i$. In terms of the labelling k, this means that the norm formula is true for k if and only if it is true for $k - m$, where $m \in \mathbb{Z}^5$ is an element of the group M generated by the six vectors in which two of the first four components are equal to 1 and the remaining three are zero. This group M consists of the vectors $(m_1, m_2, m_3, m_4, 0) \in \mathbb{Z}^5$ such that $m_1 + \cdots + m_4$ is even. Hence we reduce to the situation where $k_2 = k_3 = k_4 = 0$, i.e. to the case of $S = S(R)$ with R of type B_n, already dealt with in (a) above.

(d) We have now established the norm formula (5.8.17) for all affine root systems S, under the restriction that the labels $k(a)$ are integers ≥ 0. To remove this restriction we may argue as follows. First, in view of (5.3.13), we may assume that $\lambda = 0$, so that we are required to prove that

$$<1, 1>_k = \Delta^+_{S',k'}(\rho_{k'})\Delta^-_{S',-k'}(-\rho_{k'}) \tag{5.8.18}$$

for arbitrary k. Both sides of (5.8.18) are meromorphic functions of q, where $|q| < 1$, and $r \leq 5$ other variables $t_1, \ldots, t_r$, say (where $\{t_1, \ldots, t_r\} = \{q^{k(a)} : a \in S\}$). As we have seen, the two sides of (5.8.18) are equal whenever each t_i is a positive integral power of q. Hence to complete the proof it is enough to show that they are equal when $t_1 = \cdots = t_r = 0$, i.e. when $k(a) \to \infty$ for all $a \in S$.

From (5.1.35) we have in this situation

$$<1, 1>_\infty = (1, 1)_\infty = \mathrm{ct}(\Delta_{S,\infty})$$

and by (5.1.7)

$$\Delta_{S,\infty} = \prod_{\substack{a \in S^+ \\ 2a \notin S}} (1 - e^a).$$

From [M2], Theorem (8.1) (namely the denominator formula for affine Lie algebras), it follows that the constant term of $\Delta_{S,\infty}$ when $S = S(R)$ is $(q;q)_\infty^{-n}$ (where n is the rank of R). On the other hand, it is easily seen that the right-hand side of (5.8.18) reduces to $(q;q)_\infty^{-n}$ when $k = \infty$. Hence when $S = S(R)$, the two sides of (5.8.18) are equal at $t_1 = \cdots = t_r = 0$. Likewise, when $S = S(R)^\vee$, both sides of (5.8.18) are equal to $\prod_{i \in I_0} (q_{\alpha_i}; q_{\alpha_i})^{-1}$ when $t_1 = \cdots = t_r = 0$. This completes the proof of the norm formula (5.8.17). □

Finally, we shall calculate (E_λ, E_λ) for $\lambda \in L$. The result is

$$(E_\lambda, E_\lambda) = \prod_{a' \in S'(\lambda)} (\Delta_{a',k'} \Delta_{a',-k'})(r_{k'}(\lambda)) \tag{5.8.19}$$

where

$$S'(\lambda) = \{a' \in S'^+ : \chi(Da') + <\lambda, Da'> > 0\}.$$

In particular, when $\lambda = 0$:

$$(1, 1) = (\Delta^-_{S',k'} \Delta^-_{S',-k'})(-\rho_{k'}). \tag{5.8.20}$$

Proof First of all, (5.8.20) follows from (5.8.18) by use of (5.1.35), (5.1.40) and (5.1.41):

$$\begin{aligned}(1, 1) &= W_0(q^k)<1, 1> \\ &= \Delta^0_{S',k'}(-\rho_{k'})^{-1} \Delta^+_{S',k'}(\rho_{k'}) \Delta^-_{S',-k'}(-\rho_{k'}) \\ &= \Delta^-_{S',k'}(-\rho_{k'}) \Delta^-_{S',-k'}(-\rho_{k'}).\end{aligned}$$

Next, from (5.2.15) we have

$$(E_\lambda, E_\lambda)_1 = \prod_{a' \in S'(u'(\lambda)^{-1})} ((\Delta_{a',k'} \Delta_{a',-k'})(r_{k'}(\lambda)))^{-1}. \tag{1}$$

Let $b' = -u'(\lambda)^{-1} a' \in S'^+$. Then $a'(r_{k'}(\lambda)) = -b'(-\rho_{k'})$, so that (1) becomes

$$(E_\lambda, E_\lambda)_1 = \prod_{b' \in S'(u'(\lambda))} ((\Delta_{b',k'} \Delta_{b',-k'})(-\rho_{k'}))^{-1} \tag{2}$$

(since $\Delta_{-b',k'} \Delta_{-b',-k'} = \Delta_{b',k'} \Delta_{b',-k'}$).

If $b' \in S'(u'(\lambda))$ then $Db' < 0$ by (2.4.7) (i). Hence from (2) and (5.8.20) we obtain

$$\begin{aligned}(E_\lambda, E_\lambda) &= (E_\lambda, E_\lambda)_1 (1, 1) \\ &= \prod_{b'} (\Delta_{b',k'} \Delta_{b',-k'})(-\rho_{k'}) \end{aligned} \tag{3}$$

where the product is over $b' \in S'^+$ such that $u'(\lambda)b' \in S'^+$ and $Db' < 0$. Equivalently, with $a' = u'(\lambda)b'$,

$$(4) \qquad (E_\lambda, E_\lambda) = \prod_{a'} (\Delta_{a',k'}\Delta_{a',-k'})(r_{k'}(\lambda))$$

where the product is now over $a' \in S'^+$ such that $b' = u'(\lambda)^{-1}a' \in S'^+$ and $Db' < 0$. If $a' = \alpha' + rc$ then

$$b' = u'(\lambda)^{-1}a' = v(\lambda)\alpha' + (<\lambda, \alpha'> + r)c.$$

Now by (2.4.6) $v(\lambda)\alpha' < 0$ if and only if $<\lambda, \alpha'> + \chi(\alpha) > 0$. Hence if $a' \in S'^+$ and $v(\lambda)\alpha' < 0$ then

$$<\lambda, \alpha'> + r \geq <\lambda, \alpha'> + \chi(\alpha') > 0$$

so that $b' \in S'^+$. Hence the set of a' in the product (4) is precisely $S'(\lambda)$. □

Suppose in particular that $S = S(R)$ (1.4.1) and the labels $k(\alpha)$ are positive integers. Then $S' = S(R^\vee)$ and $\alpha^\vee + rc \in S'(\lambda)$ if and only if $r \geq \chi(\alpha)$ and $<\lambda, \alpha^\vee> + \chi(\alpha) > 0$. If $\alpha \in R^+$ and $<\lambda, \alpha^\vee> > 0$ we get a contribution

$$\prod_{i=0}^{k(\alpha)-1} \frac{1 - q^{<r_k(\lambda), \alpha^\vee> + i}}{1 - q^{<r_k(\lambda), \alpha^\vee> - i - 1}}$$

to (E_λ, E_λ). If on the other hand $\alpha \in R^+$ and $<\lambda, \alpha^\vee> \leq 0$ then $-\alpha^\vee + rc \in S'(\lambda)$ for $r \geq 1$, and we get a contribution

$$\prod_{i=0}^{k(\alpha)-1} \frac{1 - q^{-<r_k(\lambda), \alpha^\vee> + i + 1}}{1 - q^{-<r_k(\lambda), \alpha^\vee> - i}}.$$

Hence in terms of

$$[s] = q^{s/2} - q^{-s/2}$$

we obtain

$$(5.8.21) \quad (E_\lambda, E_\lambda) = q^{N(k)/2} \prod_{\alpha \in R^+} \left(\prod_{i=0}^{k(\alpha)-1} \frac{[<r_k(\lambda), \alpha^\vee> + i]}{[<r_k(\lambda), \alpha^\vee> - i - 1]} \right)^{\eta(<\lambda, \alpha^\vee>)}$$

(where $N(k) = \sum_{\alpha \in R^+} k(\alpha)^2$ as in (5.1.16)), in agreement with [M7], (7.5). (Note that C_k as defined in [M7] is equal to $q^{-N(k)/2}\Delta_k$.)

Finally, we shall indicate another method of calculating (E_λ, E_λ) where $\lambda \in L$. This method uses results from §5.5–§5.7 to express (E_λ, E_λ) in terms of

$<P_\mu, P_\mu>$, where $\mu = \lambda_+$ is the dominant weight in the orbit $W_0\lambda$. (When λ itself is dominant we have already done this, in the proof of (5.7.12).)

Recall from Chapter 2 that $v(\lambda)$ (resp. $\bar{v}(\lambda)$) is the shortest element $w \in W_0$ such that $w\lambda = w_0\mu$ (resp. $w\mu = \lambda$). Thus $v(\lambda)\bar{v}(\lambda)$ takes μ to $w_0\mu$, and $w_0 = v(\lambda)\bar{v}(\lambda)w_{0\mu}$, where $w_{0\mu}$ is the longest element of W_0 that fixes μ. We have

$$l(w_0) = l(v(\lambda)) + l(\bar{v}(\lambda)) + l(w_{0\mu})$$

and therefore

(5.8.22) $$\tau_{w_0} = \tau_{v(\lambda)}\tau_{\bar{v}(\lambda)}\tau_{w_{0\mu}}.$$

Let

$$F_\lambda = U^+E_\lambda$$

which by (5.7.2) is a scalar multiple of F_μ. In fact

$$F_\mu = \varphi_\lambda F_\lambda$$

where

(5.8.23) $$\varphi_\lambda = \prod_{a'} \Delta_{a',k'}(r_{k'}(\lambda))$$

and the product is over $a' \in S_0'^{-}$ such that $<\lambda, a'> > 0$.

Proof Let $i \in I_0$. From (5.7.2), if $<\lambda, \alpha_i'> < 0$ we have

$$F_\lambda = \tau_i c_{-\alpha_i',k'}(r_{k'}(\lambda))F_{s_i\lambda}.$$

Now by (2.7.2) (ii), if α' is positive then $\bar{v}(\lambda)^{-1}\alpha'$ is a negative root if and only if $<\lambda, \alpha'> < 0$. By taking a reduced expression for $\bar{v}(\lambda)^{-1}$, it follows that

$$F_\lambda = \Big(\tau_{\bar{v}(\lambda)} \prod_{\substack{\alpha' \in R'^{-} \\ <\lambda,\alpha'>>0}} c_{\alpha',k'}(r_{k'}(\lambda))\Big) F_\mu$$

which by (5.1.2) gives the stated value for φ_λ. □

Next we have

(5.8.24) $$(E_\lambda, E_\lambda) = \tau_{w_0}^2 W_{0\mu}(\tau^{-2}) <P_\mu, P_\mu> / \varphi_\lambda^* \xi_\lambda^{(-1)}$$

where

(5.8.25) $$\xi_\lambda^{(-1)} = \tau_{v(\lambda)}^2 \prod_{\substack{a' \in S_0'^+ \\ <\lambda, a> > 0}} \Delta_{a',-k'}(r_{k'}(\lambda))^{-1}.$$

Proof From (5.7.7) and (5.7.10) we have

$$P_\mu = \tau_{w_0} W_{0\mu}(\tau^2)^{-1} F_\mu$$

and hence

$$\begin{aligned}(P_\mu, P_\mu) &= \frac{\varphi_\lambda \varphi_\lambda^*(U^+ E_\lambda, U^+ E_\lambda)}{W_{0\mu}(\tau^2) W_{0\mu}(\tau^{-2})} \\ &= \frac{W_0(q^k)\varphi_\lambda \varphi_\lambda^*(U^+ E_\lambda, E_\lambda)}{\tau_{w_0} W_{0\mu}(\tau^2) W_{0\mu}(\tau^{-2})} \qquad \text{by (5.5.17),} \\ &= \frac{W_0(q^k)\varphi_\lambda^*(P_\mu, E_\lambda)}{\tau_{w_0}^2 W_{0\mu}(\tau^{-2})}.\end{aligned}$$

Now by (5.7.8)

$$P_\mu = \sum_{\lambda \in W_0 \mu} \xi_\lambda^{(-1)} E_\lambda$$

where (5.6.7)

$$\xi_\lambda^{(-1)} = \tau_{v(\lambda)}\, \boldsymbol{c}_{S',-k'}(v(\lambda))(r_{k'}(\lambda))$$

which agrees with the value of $\xi_\lambda^{(-1)}$ stated above. Hence we obtain

$$(P_\mu, P_\mu) = \frac{W_0(q^k)\varphi_\lambda^* \xi_\lambda^{(-1)}(E_\lambda, E_\lambda)}{\tau_{w_0}^2 W_{0\mu}(\tau^{-2})}.$$

Since

$$<P_\mu, P_\mu> = W_0(q^k)^{-1}(P_\mu, P_\mu)$$

by (5.1.35) and (5.3.2), we obtain (E_λ, E_λ) as stated. □

It remains to recast the right-hand side of (5.8.24) in the form of (5.8.19). Consider first $<P_\mu, P_\mu>$: by (5.8.17),

$$<P_\mu, P_\mu> = \prod_{\substack{a' \in S'^+ \\ Da' > 0}} \Delta_{a',k'}(\mu + \rho_{k'}) \prod_{\substack{a' \in S'^+ \\ Da' < 0}} \Delta_{a',-k'}(-\mu - \rho_{k'}).$$

This is unaltered by replacing μ by $-w_0\mu$, since $-w_0$ permutes the factors in each of the two products. We have

$$w_0\mu - \rho_{k'} = r_{k'}(w_0\mu) = r_{k'}(v(\lambda)\lambda) = v(\lambda) r_{k'}(\lambda).$$

Hence, putting $a' = \alpha' + rc$ and $b' = \beta' + rc$, where $\beta' = -v(\lambda)^{-1}\alpha'$ in the first product, and $\beta' = v(\lambda)^{-1}\alpha'$ in the second product, we shall obtain

$$(1) \qquad <P_\mu, P_\mu> = \prod_{b' \in \Sigma_0} \Delta_{b',k'}(r_{k'}(\lambda)) \prod_{b' \in \Sigma_1} \Delta_{b',-k'}(r_{k'}(\lambda))$$

where

$$\Sigma_0 = \{b' = \beta' + rc \in S' : v(\lambda)\beta' \in S_0'^- \text{ and } r \geq 0\},$$
$$\Sigma_1 = \{b' = \beta' + rc \in S' : v(\lambda)\beta' \in S_0'^- \text{ and } r > 0\}.$$

By (2.4.6), $v(\lambda)\beta' \in S_0'^-$ if and only if $\chi(\beta') + <\lambda, \beta'> > 0$. Hence

$$\Sigma_0 = S'(\lambda) \cup \{\beta' \in S_0'^- : <\lambda, \beta'> \geq 0\},$$
$$\Sigma_1 = S'(\lambda) - \{\beta' \in S_0'^+ : <\lambda, \beta'> > 0\},$$

so that (1) above becomes

$$(2) \qquad <P_\mu, P_\mu> = c_1 c_2 c_3 \prod_{a' \in S'(\lambda)} (\Delta_{a',k'}\Delta_{a',-k'})(r_{k'}(\lambda)),$$

where

$$(3) \qquad c_1 = \prod_{\substack{\beta' \in S_0'^- \\ <\lambda,\beta'>>0}} \Delta_{\beta',k'}(r_{k'}(\lambda)) = \varphi_\lambda = \tau_{\bar{v}(\lambda)}^{-2}\varphi_\lambda^*$$

by (5.8.23),

$$(4) \qquad c_2 = \prod_{\substack{\beta' \in S_0'^+ \\ <\lambda,\beta'>>0}} \Delta_{\beta',-k'}(r_{k'}(\lambda))^{-1} = \tau_{v(\lambda)}^{-2}\xi_\lambda^{(-1)}$$

by (5.8.25), and

$$c_3 = \prod_{\substack{\beta' \in S_0'^- \\ <\lambda,\beta'>=0}} \Delta_{\beta',k'}(r_{k'}(\lambda)).$$

Now if $\beta' \in S_0'^-$ and $<\lambda, \beta'> = 0$, we have

$$<r_{k'}(\lambda), \beta'> = <\lambda - v(\lambda)^{-1}\rho_{k'}, \beta'> = <\rho_{k'}, \alpha'>$$

where $\alpha' = -v(\lambda)\beta' \in S_0'^+$. Also

$$<\mu, -w_0\alpha'> = <v(\lambda)\lambda, -\alpha'> = <\lambda, \beta'> = 0$$

so that

$$(5) \qquad c_3 = \prod_{\substack{\alpha' \in S_0'^+ \\ <\mu,\alpha'>=0}} \Delta_{\alpha',k'}(\rho_{k'}) = W_{0\mu}(\tau^2)^{-1}$$

by (5.1.40).

If we now substitute (2)–(5) into the right-hand side of (5.8.24) and make use of (5.8.22), we shall finally obtain

$$(E_\lambda, E_\lambda) = \prod_{a' \in S'(\lambda)} (\Delta_{a',k'}\Delta_{a',-k'})(r_{k'}(\lambda))$$

as desired. □

5.9 Shift operators

In this section we shall give another proof of the relation (5.8.16), using shift operators. Unlike the previous proof, it makes essential use of duality (§4.7). We retain the notation of the previous sections.

For each indivisible $a' \in S'$, let

$$(5.9.1) \qquad \delta_{a'} = \delta_{a',k'} = \begin{cases} (e^{a'/2} - e^{-a'/2})\boldsymbol{c}_{a',k'} & \text{if } 2a' \notin S', \\ (e^{a'} - e^{-a'})\boldsymbol{c}_{a',k'} & \text{if } 2a' \in S', \end{cases}$$

(so that $\delta_{a'}^* = -\delta_{a'}$), and let

$$(5.9.2) \qquad \delta'_{\varepsilon,k'} = \prod_{\substack{a' \in S_{01}'^+ \\ l'(a')=1}} \delta_{a',k'},$$

where $S_{01}'^+ = \{a' \in S'^+ : a'(0) = 0 \text{ and } \frac{1}{2}a' \notin S'\}$.

(5.9.3) *For each $i \in I_0$ we have*

(i) $\left(T_i - \tau_i^{(\varepsilon)}\right)\delta'_{\varepsilon,k'}(Y^{-1}) = (s_i\delta'_{\varepsilon,k'})(Y^{-1})(T_i - \tau_i)$,

(ii) $(T_i - \tau_i)\delta'_{\varepsilon,k'}(Y) = (s_i\delta'_{\varepsilon,k'})(Y)\left(T_i - \tau_i^{(\varepsilon)}\right)$.

Proof These follow from (5.8.7) by taking adjoints (5.1.22) and then applying duality (4.7.6). □

Now let

$$G_\varepsilon = \delta_{\varepsilon,k}(X)^{-1}\delta'_{\varepsilon,k'}(Y^{-1}),$$
$$\hat{G}_\varepsilon = \delta'_{\varepsilon,k'}(Y)\delta_{\varepsilon,k}(X).$$

(5.9.4) *G_ε and $\hat{G}_\varepsilon$ each map A_0 to A_0.*

Proof Let $f \in A_0$. Then $(T_i - \tau_i)f = 0$ for all $i \neq 0$, and hence by (5.9.3)(i) $\delta'_{\varepsilon,k'}(Y^{-1})f$ is killed by $T_i - \tau_i^{(\varepsilon)}$, and hence lies in $U_\varepsilon A = \delta_{\varepsilon,k} A_0$ (5.8.8). Consequently $G_\varepsilon f \in A_0$.

Next, $\delta_{\varepsilon,k} f \in U_\varepsilon A$, hence is killed by $T_i - \tau_i^{(\varepsilon)}$, so that by (5.9.3)(ii) we have $(T_i - \tau_i)\hat{G}_\varepsilon f = 0$ for all $i \neq 0$, and therefore $\hat{G}_\varepsilon f \in A_0$. □

Next we have

(5.9.5)
$$\delta'_{\varepsilon,k'}(Y^{-1})U^+ = U_\varepsilon \delta'_{\varepsilon,k'}(Y).$$

Proof By duality it is enough to show that

$$\delta_{\varepsilon,k}(X^{-1})U_\varepsilon = U^+ \delta_{\varepsilon,k}(X).$$

By (5.8.7) we have $(T_i - \tau_i)\delta_{\varepsilon,k}(X^{-1})U_\varepsilon = 0$ for all $i \neq 0$, and hence by (5.5.10)(ii)

$$\delta_{\varepsilon,k}(X^{-1})U_\varepsilon = U^+ f(X)$$

for some $f \in A$. Now U_ε and U^+ are both of the form $T(w_0)$ + lower terms, i.e. of the form

$$\boldsymbol{c}_+(X)w_0 + \text{ lower terms},$$

hence

$$\delta_{\varepsilon,k}(X^{-1})\boldsymbol{c}_+(X) = \boldsymbol{c}_+(X)(w_0 f)(X)$$

giving $f(X) = \delta_{\varepsilon,k}(X)$ as required. □

(5.9.6) *Let $f, g \in A_0$. Then*

$$<G_\varepsilon f, g^0>_{k+l} = q^{k \cdot l} <f, (\hat{G}_\varepsilon g)^0>_k,$$

where $k \cdot l = \sum_{a \in S_{01}^+} k(a)l(a)$.

Proof By (5.1.34) and (5.8.6) we have

$$
\begin{aligned}
<G_\varepsilon f, g^0>_{k+l} &= (G_\varepsilon f, g)_{k+l} / W_0(q^{k+l}) \\
&= (\delta'_{\varepsilon,k'}(Y^{-1})f, \delta_{\varepsilon,k}(X)g)_k / W_0(q^{\varepsilon k}). \qquad (1)
\end{aligned}
$$

Since $f \in A_0$ we have

$$U^+ f = \tau_{w_0}^{-1} W_0(q^k) f$$

and therefore

$$\begin{aligned} \delta'_{\varepsilon,k'}(Y^{-1}) f &= \frac{\tau_{w_0}}{W_0(q^k)} \delta'_{\varepsilon,k'}(Y^{-1}) U^+ f \\ &= \frac{\tau_{w_0}}{W_0(q^k)} U_\varepsilon \delta'_{\varepsilon,k'}(Y) f \end{aligned}$$

by (5.9.5). Since U_ε is self-adjoint ((5.5.17)(iii)), it follows that (1) is equal to

(2) $$\tau_{w_0}(\delta'_{\varepsilon,k'}(Y) f, U_\varepsilon \delta_{\varepsilon,k}(X) g)_k / W_0(q^k) W_0(q^{\varepsilon k}).$$

Now $\delta_{\varepsilon,k}(X) g \in U_\varepsilon A$ by (5.8.8), hence by (5.5.17)(ii)

$$U_\varepsilon \delta_{\varepsilon,k}(X) g = \left(\tau_{w_0}^{(\varepsilon)}\right)^{-1} W_0(q^{\varepsilon k}) \delta_{\varepsilon,k}(X) g.$$

Since $\tau_{w_0}/\tau_{w_0}^{(\varepsilon)} = \varepsilon(w_0) q^{k \cdot l}$, it follows that (2) is equal to

$$\varepsilon(w_0) q^{k \cdot l} (\delta'_{\varepsilon,k'}(Y) f, \delta_{\varepsilon,k}(X) g)_k / W_0(q^k)$$

which in turn is equal to

$$q^{k \cdot l}(f, \hat{G}_\varepsilon g)_k / W_0(q^k) = q^{k \cdot l} < f, (\hat{G}_\varepsilon g)^0 >_k$$

since the adjoint of $\delta'_{\varepsilon,k'}(Y)$ is $\varepsilon(w_0) \delta'_{\varepsilon,k'}(Y)$ by (5.1.24). □

(5.9.7) *Let* $\lambda \in L_{++}$. *Then*

$$\begin{aligned} G_\varepsilon P_{\lambda+\rho_l,k} &= d_{k,l}(\lambda) P_{\lambda,k+l}, \\ \hat{G}_\varepsilon P_{\lambda,k+l} &= \hat{d}_{k,l}(\lambda) P_{\lambda+\rho_l,k}, \end{aligned}$$

where

$$\begin{aligned} d_{k,l}(\lambda) &= q^{k \cdot l/2} \delta'_{\varepsilon,-k'}(\lambda + \rho_{k'+l'}), \\ \hat{d}_{k,l}(\lambda) &= \varepsilon(w_0) q^{-k \cdot l/2} \delta'_{\varepsilon,k'}(\lambda + \rho_{k'+l'}). \end{aligned}$$

Proof Let $\mu \in L_{++}$, $\mu < \lambda$. By (5.9.6) we have

$$< G_\varepsilon P_{\lambda+\rho_l,k}, m_\mu >_{k+l} = q^{k \cdot l} < P_{\lambda+\rho_l,k}, (\hat{G}_\varepsilon m_\mu)^0 >_k.$$

Now the leading monomial in $\delta_{\varepsilon,k} m_\mu$ is $e^{w_0(\mu+\rho_l)}$, and therefore $(\hat{G}_\varepsilon m_\mu)^0$ is a scalar multiple of $m_{\mu+\rho_l}$ + lower terms. It follows that $< G_\varepsilon P_{\lambda+\rho_l,k}, m_\mu >_{k+l} = 0$ for all $\mu \in L_{++}$ such that $\mu < \lambda$, and hence that $G_\varepsilon P_{\lambda+\rho_l,k}$ is a scalar multiple of $P_{\lambda,k+l}$, say

$$G_\varepsilon P_{\lambda+\rho_l,k} = d_{k,l}(\lambda) P_{\lambda,k+l}.$$

Hence

$$\delta'_{\varepsilon,k'}(Y^{-1})P_{\lambda+\rho_l,k} = d_{k,l}(\lambda)\delta_{\varepsilon,k}P_{\lambda,k+l}. \tag{1}$$

Since $P_{\lambda+\rho_l,k} = E_{w_0(\lambda+\rho_l),k}$ + lower terms, it follows from (5.2.2) that the coefficient of $e^{w_0(\lambda+\rho_l)}$ in the left-hand side of (1) is

$$\begin{aligned}\delta'_{\varepsilon,k'}(r_{k'}(w_0(\lambda+\rho_l)) &= \delta'_{\varepsilon,k'}(w_0(\lambda+\rho_{k'+l'})) \\ &= \varepsilon(w_0)\delta'_{\varepsilon,-k'}(\lambda+\rho_{k'+l'})\end{aligned}$$

(note that $l' = l$ in all cases); whereas on the right-hand side of (1) the coefficient is

$$d_{k,l}(\lambda)\varepsilon(w_0)q^{-k\cdot l/2}.$$

This gives the stated value for $d_{k,l}(\lambda)$. For $\hat{G}_\varepsilon$, the proof is analogous and is left to the reader. □

In view of (5.9.7), the operators G_ε and $\hat{G}_\varepsilon$ are called *shift operators*: G_ε shifts the labelling k upwards to $k+l$, and $\hat{G}_\varepsilon$ shifts down from $k+l$ to k.

From (5.9.6) and (5.9.7) we deduce

$$\frac{|P_{\lambda,k+l}|^2_{k+l}}{|P_{\lambda+\rho_l,k}|^2_k} = q^{k\cdot l}\frac{\delta'_{\varepsilon,k'}(\lambda+\rho_{k'+l'})}{\delta'_{\varepsilon,-k'}(\lambda+\rho_{k'+l'})}. \tag{5.9.8}$$

Proof Take $f = P_{\lambda+\rho_l,k}$ and $g = P_{\lambda,k+l}$ in (5.9.6). By (5.9.7) we have

$$<G_\varepsilon f, g^0>_{k+l} = d_{k,l}(\lambda)|P_{\lambda,k+l}|^2_{k+l}$$

and

$$<f, (\hat{G}_\varepsilon g)^0>_k = \hat{d}_{k,l}(\lambda)^0|P_{\lambda+\rho_l,k}|^2_k.$$

Hence

$$\frac{|P_{\lambda,k+l}|^2_{k+l}}{|P_{\lambda+\rho_l,k}|^2_k} = q^{k\cdot l}\frac{\hat{d}_{k,l}(\lambda)^0}{d_{k,l}(\lambda)}$$

which gives (5.9.8). □

To reconcile (5.9.8) with (5.8.16), suppose for example that $S = S(R)$ (1.4.1); then $S' = S(R^\vee)$ and $k' = k$, so that

$$\Delta^+_{S',k+l}/\Delta^+_{S',k} = \prod_{\substack{\alpha\in R^+\\ l(\alpha)=1}} (1-q^{k(\alpha)}e^{\alpha^\vee})$$

and

$$\Delta^{-}_{S',-k-l}/\Delta^{-}_{S',-k} = \prod_{\substack{\alpha\in R^+\\ l(\alpha)=1}} \left(1-q^{-k(\alpha)}e^{-\alpha^\vee}\right)^{-1}.$$

Hence the right-hand side of (5.8.16) is equal to

(1) $$\prod_{\alpha\in R^+} \frac{1-q^{k(\alpha)+<\lambda+\rho_{k+l},\alpha^\vee>}}{1-q^{-k(\alpha)+<\lambda+\rho_{k+l},\alpha^\vee>}}.$$

On the other hand,

$$\begin{aligned} q^{k\cdot l}\delta'_{\varepsilon,k'}/\delta'_{\varepsilon,-k'} &= q^{k\cdot l}\prod_{\substack{\alpha\in R^+\\ l(\alpha)=1}} \frac{q^{k(\alpha)/2}e^{\alpha^\vee/2}-q^{-k(\alpha)/2}e^{-\alpha^\vee/2}}{q^{-k(\alpha)/2}e^{\alpha^\vee/2}-q^{k(\alpha)/2}e^{-\alpha^\vee/2}} \\ &= \prod_{\substack{\alpha\in R^+\\ l(\alpha)=1}} \frac{1-q^{k(\alpha)}e^{\alpha^\vee}}{1-q^{-k(\alpha)}e^{\alpha^\vee}}, \end{aligned}$$

and therefore the right-hand side of (5.9.8) is equal to (1). Similarly in the other cases (1.4.2), (1.4.3).

5.10 Creation operators

The group W acts on V as a group of displacements, and by transposition acts also on F, the space of affine-linear functions on $V : (wf)(x) = f(w^{-1}x)$ for $w \in W$, $f \in F$ and $x \in V$. Since we identify $\lambda \in L$ with the function $x \mapsto <\lambda, x>$ on V, we have to distinguish $w\lambda \in V$ and $w\cdot\lambda : x \mapsto <\lambda, w^{-1}x>$. When $w \in W_0$ we have $w\lambda = w\cdot\lambda$, but for example $s_0\lambda$ and $s_0\cdot\lambda$ are not the same: we have

(5.10.1) $$\begin{aligned} s_0\lambda &= \xi + s_\xi\lambda, \\ s_0\cdot\lambda &= s_\xi\lambda + <\lambda,\xi>c \end{aligned}$$

where $\xi = \varphi$ (the highest root of R) in cases (1.4.1) and (1.4.2), and $\xi = \varepsilon_1$ in case (1.4.3).

From (4.7.3) we have

(5.10.2) $$(T_i - \boldsymbol{b}_i(X^{a_i}))X^\lambda = X^{s_i\cdot\lambda}(T_i - \boldsymbol{b}_i(X^{a_i}))$$

for all $i \in I$ and $\lambda \in L$, where

$$\boldsymbol{b}_i(X^{a_i}) = \boldsymbol{b}(\tau_i, \tau'_i; X^{a_i})$$

and in particular

$$X^{a_0} = q^m X^{-\xi}$$

where $m = \frac{1}{2}$ in case (1.4.3), and $m = 1$ otherwise.

By applying $\omega^{-1} : \tilde{\mathfrak{H}} \to \tilde{\mathfrak{H}}'$ to (5.10.2) we shall obtain

$$Y^{-\lambda}(T_i - \boldsymbol{b}_i(Y^{-a_i})) = (T_i - \boldsymbol{b}_i(Y^{-a_i}))Y^{-s_i\lambda}$$

when $i \neq 0$, and

$$Y^{-\lambda}(\omega^{-1}(T_0) - \boldsymbol{b}_0(q^m Y^{\xi})) = q^{<\lambda,\xi>}(\omega^{-1}(T_0) - \boldsymbol{b}_0(q^m Y^{\xi}))Y^{-s_\xi\lambda}.$$

Hence if we define

$$\alpha_i = T_i - \boldsymbol{b}_i(Y^{-a_i}) \tag{5.10.3}$$

for $i \neq 0$, and

$$\alpha_0 = \omega^{-1}(T_0) - \boldsymbol{b}_0(q^m Y^{\xi}) \tag{5.10.4}$$

as operators on A', we shall have

$$Y^{\lambda}\alpha_i = \alpha_i Y^{s_i\lambda} \tag{5.10.5}$$

for $i \neq 0$ and $\lambda \in L$, and

$$Y^{\lambda}\alpha_0 = q^{-<\lambda,\xi>}\alpha_0 Y^{s_\xi\lambda}. \tag{5.10.6}$$

Suppose first that $i \neq 0$, and let $\mu \in L'$. Then we have

$$Y^{\lambda}\alpha_i E'_\mu = \alpha_i Y^{s_i\lambda} E'_\mu = q^{-<s_i\lambda, r'_k(\mu)>}\alpha_i E'_\mu$$

by (5.2.2′). Suppose that $s_i\mu > \mu$, then $s_i(r'_k(\mu)) = r'_k(s_i\mu)$ by (2.8.4), and hence $\alpha_i E'_\mu$ is a scalar multiple of $E'_{s_i\mu}$. To obtain the scalar, we need the coefficient of $e^{s_i\mu}$ in $\alpha_i E'_\mu$. Now $\boldsymbol{b}_i(Y^{-a_i})E'_\mu$ is a scalar multiple of E'_μ, hence does not contain $e^{s_i\mu}$. Since $s_i\mu > \mu$ we have $<\mu, \alpha_i> > 0$ by (2.7.9) and hence

$$T_i e^{\mu} = \tau_i^{-1} e^{s_i\mu} + \text{lower terms}$$

by (4.3.21). It follows that

$$\alpha_i E'_\mu = \tau_i^{-1} E'_{s_i\mu} \tag{5.10.7}$$

if $i \neq 0$ and $s_i\mu > \mu$.

Next, consider the case $i = 0$. Then we have, using (5.10.6),

$$\begin{aligned} Y^{\lambda}\alpha_0 E'_\mu &= q^{-<\lambda,\xi>}\alpha_0 Y^{s_\xi\lambda} E'_\mu \\ &= q^{-<\lambda,\xi>-<s_\xi\lambda, r'_k(\mu)>}\alpha_0 E'_\mu \\ &= q^{-<\lambda, s_0(r'_k\mu)>} E'_\mu \end{aligned}$$

by (5.10.1). Suppose that $s_0\mu > \mu$, then $s_0(r'_k\mu) = r'_k(s_0\mu)$ by (2.7.13), and hence $\boldsymbol{\alpha}_0 E'_\mu$ is a scalar multiple of $E'_{s_0\mu}$. We shall show that in fact

(5.10.8)
$$\boldsymbol{\alpha}_0 E'_\mu = \tau_{v(s_0\mu)}\tau_{v(\mu)}^{-1} E'_{s_0\mu}.$$

Proof Since $s_0 s_\xi = t(\xi)$ we have $T_0 T(s_\xi) = Y^\xi$ and therefore $T(s_\xi)\omega^{-1}(T_0) = X^{-\xi}$, so that

$$\omega^{-1}(T_0) = T(s_\xi)^{-1} X^{-\xi}.$$

As before, we require the coefficient of $e^{s_0\mu}$ in $T(s_\xi)^{-1}X^{-\xi}e^\mu = T(s_\xi)^{-1}e^{\mu-\xi}$, which by (4.3.23) is $q^{f(\mu)}$, where

$$f(\mu) = \frac{1}{2}\sum_{\alpha\in R^+} \eta(<\xi - \mu, \alpha>)\chi(s_\xi\alpha)\kappa_\alpha.$$

Now if $\alpha \in R^+$, we have $s_\xi\alpha \in R^-$ unless $<\xi, \alpha> = 0$. Hence

(1)
$$f(\mu) = \frac{1}{2}\sum_{\substack{\alpha\in R^+\\ <\xi,\alpha>>0}} \eta(<\xi - \mu, \alpha>)\kappa_\alpha.$$

On the other hand, by (4.3.25),

$$\tau_{v(\mu)} = q^{g(\mu)}$$

where

$$g(\mu) = \frac{1}{4}\sum_{\alpha\in R^+}(1 + \eta(<\mu, \alpha>))\kappa_\alpha.$$

Now if $<\xi, \alpha> = 0$ we have $<s_0\mu, \alpha> = <\mu, \alpha>$. Hence

$$g(\mu) - g(s_0\mu) = \frac{1}{4}\sum_{\substack{\alpha\in R^+\\ <\xi,\alpha>>0}} (\eta(<\mu, \alpha>) - \eta(<s_0\mu, \alpha>))\kappa_\alpha.$$

In this sum we may replace $<s_0\mu, \alpha>$ by

$$-<s_0\mu, s_\xi\alpha> = <\xi - \mu, \alpha>$$

and hence using (1) we obtain

$$f(\mu) + g(\mu) - g(s_0\mu) = \frac{1}{4}\sum_{\substack{\alpha\in R^+\\ <\xi,\alpha>>0}} (\eta(<\xi - \mu, \alpha>) + \eta(<\mu, \alpha>))\kappa_\alpha.$$

In this sum, if $\alpha \neq \xi^\vee$ then $<\xi, \alpha> = 1$, and

$$\eta(1 - <\mu, \alpha>) + \eta(<\mu, \alpha>) = 0.$$

Finally, since $s_0\mu > \mu$ we have $a_0(\mu) > 0$ and hence $<\mu, \xi> \le 0$, so that

$$\begin{aligned}\eta(<\xi - \mu, \xi^\vee>) + \eta(<\mu, \xi^\vee>) &= \eta(2 - <\mu, \xi^\vee>) + \eta(<\mu, \xi^\vee>)\\ &= 1 - 1 = 0.\end{aligned}$$

It follows that

$$f(\mu) = g(s_0\mu) - g(\mu)$$

which completes the proof. □

Next, let $j \in J$ and let

(5.10.9) $$\beta_j = \omega^{-1}(U_j^{-1}).$$

Let $\lambda \in L$. Then by (3.4.5) we have

$$U_j^{-1} X^\lambda U_j = X^{u_j^{-1} \cdot \lambda}$$

where $u_j = u(\pi_j') = t(\pi_j')v_j^{-1}$, so that

(5.10.10) $$\begin{aligned}u_j\lambda &= \pi_j' + v_j^{-1}\lambda,\\ u_j^{-1} \cdot \lambda &= v_j\lambda + <\lambda, \pi_j'>c.\end{aligned}$$

Hence

$$U_j^{-1} X^\lambda = q^{<\lambda, \pi_j'>} X^{v_j\lambda} U_j^{-1}$$

and therefore (with λ replaced by $-\lambda$)

(5.10.11) $$Y^\lambda \beta_j = q^{-<\lambda, \pi_j'>} \beta_j Y^{v_j\lambda}$$

Now let $\mu \in L'$. Then

$$\begin{aligned}Y^\lambda \beta_j E_\mu' &= q^{-<\lambda, \pi_j'>} \beta_j Y^{v_j\lambda} E_\mu'\\ &= q^{-<\lambda, \pi_j' + v_j^{-1} r_k'(\mu)>} \beta_j E_\mu'\\ &= q^{-<\lambda, u_j(r_k'(\mu))>} \beta_j E_\mu'.\end{aligned}$$

Since $u_j(r_k'(\mu)) = r_k'(u_j\mu)$ by (2.8.4), it follows that $\beta_j E_\mu'$ is a scalar multiple of $E_{u_j\mu}'$. In fact we have

(5.10.12) $$\beta_j E_\mu' = \tau_{v(u_j\mu)} \tau_{v(\mu)}^{-1} E_{u_j\mu}'.$$

Proof We have $U_j^{-1} = U_i$, where $i = -j$ in the notation of §2.5. Since $u_i v_i = \tau(\pi_i')$, it follows that

$$U_i = T(u_i) = Y^{\pi_i'} T(v_i)^{-1}$$

and therefore

$$\beta_j = \omega^{-1}(U_i) = T\left(v_i^{-1}\right)^{-1} X^{-\pi_i'}$$

so that

$$\beta_j e^{\mu} = T\left(v_i^{-1}\right)^{-1} e^{\mu-\pi_i'}.$$

Since $v_i^{-1}(\mu - \pi_i') = u_i^{-1}\mu = u_j\mu$, it follows from (4.3.23) that

$$\beta_j e^{\mu} = q^{f(\mu)} e^{u_j\mu} + \text{lower terms}$$

where now

$$f(\mu) = \frac{1}{2}\sum_{\alpha\in R^+} \eta(<\pi_i' - \mu, \alpha>)\chi(v_i\alpha)\kappa_\alpha.$$

By (4.2.4), $\chi(v_i\alpha) = 1$ if and only if $<\pi_i', \alpha>>0$. Now π_i' is a minuscule fundamental weight, so that $<\pi_i', \alpha> = 0$ or 1 for each $\alpha \in R^+$. Hence

(1) $$f(\mu) = -\frac{1}{2} \sum_{\substack{\alpha\in R^+ \\ <\pi_i', \alpha>=1}} \eta(<\mu, \alpha>)\kappa_\alpha$$

since $\eta(<\pi_i' - \mu, \alpha>) = \eta(1 - <\mu, \alpha>) = -\eta(<\mu, \alpha>)$.

On the other hand, by (4.3.25), we have

$$\tau_{v(\mu)} = q^{g(\mu)}$$

where

$$g(\mu) = \frac{1}{4}\sum_{\alpha\in R^+} (1 + \eta(<\mu, \alpha>))\kappa_\alpha$$

so that

(2) $$g(\mu) - g(u_j\mu) = \frac{1}{4}\sum_{\alpha\in R^+} (\eta(<\mu, \alpha>) - \eta(<u_j\mu, \alpha>)\kappa_\alpha.$$

If $<\pi_j', \alpha >= 0$ let $\beta = v_j\alpha \in R^+$. Then

$$<u_j\mu, \alpha> = <\pi_j' + v_j^{-1}\mu, \alpha> = <\mu, \beta>$$

and

$$<\pi_i', \beta> = <v_j^{-1}\pi_i', \alpha> = -<\pi_j', \alpha> = 0$$

by (2.5.9).

If on the other hand $<\pi'_j, \alpha> = 1$, let $\beta = -v_j\alpha \in R^+$. Then

$$<u_j\mu, \alpha> = 1 - <\mu, \beta>$$

so that $\eta(<u_j\mu, \alpha>) = -\eta(<\mu, \beta>)$, and $<\pi'_i, \beta> = 1$.

Hence if we define

$$\varepsilon_\beta = \begin{cases} 1 & \text{if } <\pi'_i, \beta> = 0 \\ -1 & \text{if } <\pi'_i, \beta> = 1 \end{cases}$$

we have

(3) $$\sum_{\alpha \in R^+} \eta(<u_j\mu, \alpha>)\kappa_\alpha = \sum_{\beta \in R^+} \varepsilon_\beta \eta(<\mu, \beta>)\kappa_\beta.$$

From (1), (2) and (3) it follows that

$$\begin{aligned} f(\mu) + g(\mu) - g(u_j\mu) &= \sum_{\alpha \in R^+} ((\varepsilon_\alpha - 1) + 1 - \varepsilon_\alpha)\eta(<\mu, \alpha>)\kappa_\alpha \\ &= 0. \end{aligned}$$

This completes the proof of (5.10.12). □

(5.10.13) *Let* $\mu \in L'$ *and let*

$$u(\mu) = u_j s_{i_1} \cdots s_{i_p}$$

be a reduced expression. Then

$$E'_\mu = \tau^{-1}_{v(\mu)} \boldsymbol{\beta}_j \boldsymbol{\alpha}_{i_1} \cdots \boldsymbol{\alpha}_{i_p}(1).$$

Proof For each $i \in I$, if $a_i(\mu) > 0$ then $s_i u(\mu) = u(s_i\mu) > u(\mu)$, by (2.4.14). Also, if $i \neq 0$, we have $v(\mu)s_i = v(s_i\mu) < v(\mu)$, so that $\tau_{v(s_i\mu)} = \tau_i^{-1}\tau_{v(\mu)}$. Hence (5.10.13) follows from (5.10.7), (5.10.8) and (5.10.12). □

For this reason the operators $\boldsymbol{\alpha}_i$ $(i \in I)$ and $\boldsymbol{\beta}_j$ $(j \in J)$ are called 'creation operators': they enable us to construct each E'_μ from $E'_0 = 1$. Dually, by interchanging S and S', k and k', we may define operators $\boldsymbol{\alpha}'_i$, $\boldsymbol{\beta}'_j$ on A which enable us to construct each E_λ $(\lambda \in L)$ from $E_0 = 1$.

Notes and references

The symmetric scalar product (5.1.29) was introduced in [M5], and the non-symmetric scalar product (5.1.17) (which is more appropriate in the context of the action of the double affine Hecke algebra) by Cherednik [C2]. The polynomials E_λ were first defined by Opdam [O4] in the limiting case $q \to 1$, and then for arbitrary q in [M7] (for the affine root systems $S(R)$ with R reduced), and in greater generality by Cherednik in [C3]. The proofs of the symmetry and evaluation theorems in §5.2 and §5.3 are due to Cherednik ([C4], [C5]), as is indeed the greater part of the material in this chapter.

To go back in time a bit, the symmetric polynomials P_λ were first developed for the root systems of type A_n in the 1980's, as a common generalization of the Hall-Littlewood and Jack symmetric functions [M6]. The symmetry theorem (5.3.5) in this case was discovered by Koornwinder, and his proof is reproduced in [M6], Chapter 6, which also contains the evaluation theorem (5.3.12) and the norm formula (5.8.17) for S of type A_n, without any overt use of Hecke algebras. (Earlier, Stanley [S3] had done this in the limiting case $q \to 1$, i.e. for the Jack symmetric functions.) What is special to the root systems of type A is that all the fundamental weights are minuscule, so that the corresponding Y-operators (4.4.12) can be written down explicitly; and these are precisely the operators used in [M6].

From the nature of these formulas in type A it was clear what to expect should happen for other root systems – all the more because the formula for $|P_\lambda|^2$ when $\lambda = 0$ delivers the constant term of ∇, which had been the subject of earlier conjectures ([D1], [A1], [M4], [M11]). The preprint [M5] contained a construction of the polynomials P_λ for reduced affine root systems, and conjectured the values of $|P_\lambda|^2$ and $P_\lambda(\rho'_k)$. Again, in the limiting case $q \to 1$, Heckman and Opdam ([H1], [H2], [O1], [O2]) had earlier constructed the P_λ (which they called Jacobi polynomials); and then Opdam [O3] saw how to exploit the shift operator techniques that he and Heckman had developed, to establish the norm and evaluation formulas in this limiting case.

Cherednik [C2] now brought the double affine Hecke algebra into the picture, as a ring of operators on A, as described in Chapter 4. He constructed q-analogues of the shift operators, and used them to evaluate $|P_\lambda|^2$ for reduced affine root systems. His proof is reproduced in §5.9. The alternative proof of the norm formula in §5.8 is essentially that of [M7], which in turn was inspired by [O4].

Finally, the case where the affine root system is of type $(C_n^\vee, C_n)$ was worked out by van Diejen [V2] in the self-dual situation (i.e., $k' = k$ in our notation), and then in general by Noumi [N1], Sahi ([S1], [S2]), and Stokman [S4]. The constant term of $\nabla_{S,k}$ (i.e., the case $\lambda = 0$ of the norm formula) had been calculated earlier by Gustafson [G2].

6

The rank 1 case

When the affine root system S has rank 1, everything can be made completely explicit, and we are dealing with orthogonal polynomials in one variable. There are two cases to consider:

(a) $S = S'$ is of type A_1, and $L = L'$ is the weight lattice;
(b) $S = S'$ is of type $(C_1^\vee, C_1)$, and $L = L'$ is the root lattice.

We consider (a) first.

6.1 Braid group and Hecke algebra (type A_1)

Here $R = R' = \{\pm\alpha\}$, where $|\alpha|^2 = 2$, and $L = L' = \mathbb{Z}\alpha/2$. We have $a_0 = 1 - \alpha$ and $a_1 = \alpha$, acting on $V = \mathbb{R}$ as follows: $a_0(\xi) = 1 - \xi$ and $a_1(\xi) = \xi$ for $\xi \in \mathbb{R}$. Thus the simplex C is the interval (0, 1), and W_S is the infinite dihedral group, freely generated by s_0 and s_1, where s_0 (resp. s_1) is reflection in 1 (resp. 0). The extended affine Weyl group W is the extension of W_S by a group $\Omega = \{1, u\}$ of order 2, where u is reflection in the point $\frac{1}{2}$, so that u interchanges 0 and 1, a_0 and a_1. We have $s_0 = us_1u$, so that W is generated by s_1 and u with the relations $s_1^2 = u^2 = 1$.

The *braid group* $\mathfrak{B}$ has generators T_0, T_1, U with relations

$$U^2 = 1, \quad UT_1U = T_0 \tag{6.1.1}$$

(there are no braid relations). Let $Y = Y^{\alpha/2}$, then

$$Y = T_0U = UT_1$$

so that $\mathfrak{B}$ is generated by T_1 and Y subject to the relation $T_1Y^{-1}T_1 = Y$. Alternatively, $\mathfrak{B}$ is generated by T_1 and U with the single relation $U^2 = 1$.

The *double braid group* $\tilde{\mathfrak{B}}$ is generated by T_1, X, Y and a central element $q^{1/2}$, with the relations

(6.1.2) $$T_1 X T_1 = X^{-1}, \quad T_1 Y^{-1} T_1 = Y, \quad U X U^{-1} = q^{1/2} X^{-1}$$

where $U = Y T_1^{-1} = T_1 Y^{-1}$. The duality antiautomorphism ω maps $T_1, X, Y, q^{1/2}$ respectively to $T_1, Y^{-1}, X^{-1}, q^{1/2}$. Thus it interchanges the first two of the relations (6.1.2); and since $UX = T_1 Y^{-1} X$ we have $\omega(UX) = Y^{-1} X T_1 = T_1^{-1}(UX)T_1$, so that $\omega((UX)^2) = T_1^{-1}(UX)^2 T_1 = q^{1/2}$. Thus duality is directly verified.

Next, the *affine Hecke algebra* $\mathfrak{H}$ is the K-algebra generated by T_1 and U subject to the relations $U^2 = 1$ and

(6.1.3) $$(T_1 - \tau)(T_1 + \tau^{-1}) = 0$$

where K is the field $\mathbb{Q}(q^{1/2}, \tau)$. We shall write

$$\tau = q^{k/2}$$

and we shall assume when convenient that k is a non-negative integer (in which case $K = \mathbb{Q}(q^{1/2})$).

The *double affine Hecke algebra* $\tilde{\mathfrak{H}}$ is the K-algebra generated by T_1, X, Y subject to the relations (6.1.2) and (6.1.3), i.e. it is the quotient of the group algebra of $\tilde{\mathfrak{B}}$ over K by the ideal generated by $(T_1 - \tau)(T_1 + \tau^{-1})$. Since ω fixes T_1, it extends to an antiautomorphism of $\tilde{\mathfrak{H}}$.

Let $x = e^{\alpha/2}$ and let $A = K[x, x^{-1}]$. Also let

(6.1.4) $$\boldsymbol{b}(X) = \frac{\tau - \tau^{-1}}{1 - X^2}, \quad \boldsymbol{c}(X) = \frac{\tau X^2 - \tau^{-1}}{X^2 - 1}.$$

Then $\tilde{\mathfrak{H}}$ acts on A as follows: if $f \in A$,

(6.1.5) $$Xf = xf, \quad Uf = uf, \quad T_1 f = (\boldsymbol{b}(X) + \boldsymbol{c}(X)s_1)f$$

where $(s_1 f)(x) = f(x^{-1})$ and $(uf)(x) = f(q^{1/2}x^{-1})$.

We have $s_1 X = X^{-1} s_1$, and

$$s_1 = \boldsymbol{c}(X)^{-1}(T_1 - \boldsymbol{b}(X))$$

as operators on A. Applying ω, we obtain

$$Y^{-1}(T_1 - \boldsymbol{b}(Y^{-1}))\,\boldsymbol{c}(Y^{-1})^{-1} = (T_1 - \boldsymbol{b}(Y^{-1}))\,\boldsymbol{c}(Y^{-1})^{-1} Y$$

so that if we put

(6.1.6) $$\alpha = T_1 - \boldsymbol{b}(Y^{-1}) = UY - \boldsymbol{b}(Y^{-1}),$$

we have

(6.1.7) $$Y^{-1}\alpha = \alpha Y.$$

Again, we have $UX = q^{1/2}X^{-1}U$ and $U = T_1 Y^{-1}$, so that

(6.1.8) $$\beta = \omega(U) = XT_1 = XUY$$

and

(6.1.9) $$Y^{-1}\beta = q^{1/2}\beta Y.$$

6.2 The polynomials E_m

As in §5.1, the scalar product on A is

$$(f, g) = \mathrm{ct}(fg^*\Delta_k)$$

where

$$\Delta_k = (x^2; q)_k(qx^{-2}; q)_k.$$

We shall assume that k is a non-negative integer. Then

$$\Delta_k = (-1)^k q^{k(k+1)/2} x^{-2k}(q^{-k}x^2; q)_{2k}$$

which by the q-binomial theorem is equal to

$$\sum_{r=-k}^{k} (-1)^r q^{r(r-1)/2} \begin{bmatrix} 2k \\ k+r \end{bmatrix} x^{2r}$$

where

$$\begin{bmatrix} n \\ r \end{bmatrix} = (q; q)_n/(q; q)_r(q; q)_{n-r}$$

is the q-binomial coefficient, for $0 \le r \le n$. In particular, the constant term of Δ_k is $\left[\begin{smallmatrix} 2k \\ k \end{smallmatrix}\right]$, i.e.,

(6.2.1) $$(1, 1) = \begin{bmatrix} 2k \\ k \end{bmatrix}.$$

For each $m \in \mathbb{Z}$, let $E_m = E_{m\alpha/2}$ in the notation of §5.2. We have $\rho_{k'} = \frac{1}{2}k\alpha$ and hence

$$r_{k'}(\tfrac{1}{2}m\alpha) = \begin{cases} \frac{1}{2}(m+k)\alpha & \text{if } m > 0, \\ \frac{1}{2}(m-k)\alpha & \text{if } m \le 0, \end{cases}$$

so that by (5.2.2) the E_m are elements of A characterized by the facts that the coefficient of x^m in E_m is 1, and that

(6.2.2) $$YE_m = \begin{cases} q^{-(m+k)/2}E_m & \text{if } m > 0, \\ q^{(-m+k)/2}E_m & \text{if } m \le 0. \end{cases}$$

The adjoint of Y (for the scalar product (f, g)) is Y^{-1}, from which it follows that the E_m are pairwise orthogonal. If $m > 0$, E_m is a linear combination of x^{m-2i} for $0 \le i \le m-1$, and E_{-m} is a linear combination of x^{m-2i} for $0 \le i \le m$.

(6.2.3) *If $m \ge 0$ we have*

$$E_{-m-1} = q^{k/2}\alpha E_{m+1},$$
$$E_{m+1} = q^{-k/2}\beta E_{-m}$$

where α, β *are given by* (6.1.6) *and* (6.1.8).

Proof By (6.1.7),

$$Y\alpha E_{m+1} = \alpha Y^{-1}E_{m+1} = q^{(m+k+1)/2}\alpha E_{m+1},$$

so that by (6.2.2) αE_{m+1} is a scalar multiple of E_{-m-1}. But

$$\alpha E_{m+1} = UYE_{m+1} - \boldsymbol{b}(Y^{-1})E_{m+1},$$

in which $\boldsymbol{b}(Y^{-1})E_{m+1}$ is a scalar multiple of E_{m+1}, hence does not contain x^{-m-1}. Also in

$$UYE_{m+1} = q^{-(m+k+1)/2}E_{m+1}(q^{1/2}x^{-1})$$

the coefficient of x^{-m-1} is $q^{-k/2}$. It follows that $\alpha E_{m+1} = q^{-k/2}E_{-m-1}$, which gives the first of the relations (6.2.3).

Next, by (6.1.9),

$$Y\beta E_{-m} = q^{-1/2}\beta Y^{-1}E_{-m} = q^{-(m+k+1)/2}\beta E_{-m},$$

so that by (6.2.2) βE_{-m} is a scalar multiple of E_{m+1}. But

$$\beta E_{-m} = XUYE_{-m} = q^{(m+k)/2}xE_{-m}(q^{1/2}x^{-1}),$$

in which the coefficient of x^{m+1} is $q^{k/2}$. Hence $\beta E_{-m} = q^{k/2}E_{m+1}$. □

The operators α and β are 'creation operators': from (6.2.3) we have, for $m \ge 0$,

(6.2.4) $$E_{-m} = (\alpha\beta)^m(1), \quad E_{m+1} = q^{-k/2}\beta(\alpha\beta)^m(1),$$

since $E_0 = 1$.

We shall now calculate the polynomials E_m explicitly. For this purpose we introduce

$$f(x,z) = 1/(xz;q)_k(x^{-1}z;q)_{k+1} = \sum_{m\geq 0} f_m(x)z^m,$$

$$g(x,z) = x/(xz;q)_{k+1}(qx^{-1}z;q)_k = \sum_{m\geq 0} g_m(x)z^m.$$

By the q-binomial theorem we have

$$f(x,z) = \sum_{i,j\geq 0} \begin{bmatrix} k+i-1 \\ i \end{bmatrix} \begin{bmatrix} k+j \\ j \end{bmatrix} x^{i-j}z^{i+j},$$

$$g(x,z) = \sum_{i,j\geq 0} \begin{bmatrix} k+i-1 \\ i \end{bmatrix} \begin{bmatrix} k+j \\ j \end{bmatrix} q^i x^{-i+j+1}z^{i+j},$$

so that

(6.2.5) $$f_m(x) = \sum_{i+j=m} \begin{bmatrix} k+i-1 \\ i \end{bmatrix} \begin{bmatrix} k+j \\ j \end{bmatrix} x^{i-j},$$

(6.2.6) $$g_m(x) = \sum_{i+j=m} \begin{bmatrix} k+i-1 \\ i \end{bmatrix} \begin{bmatrix} k+j \\ j \end{bmatrix} q^i x^{-i+j+1}.$$

Since $T_1 = \boldsymbol{b}(X) + \boldsymbol{c}(X)s_1$ (6.1.5), a brief calculation gives

$$T_1 f(x,z) = q^{k/2} f(q^{1/2}x^{-1}, q^{1/2}z),$$

$$T_1 g(x,z) = q^{-(k+1)/2} g(q^{1/2}x^{-1}, q^{1/2}z).$$

Since $Y = UT_1$ it follows that

$$Yf(x,z) = q^{k/2} f(x, q^{1/2}z),$$

$$Yg(x,z) = q^{-(k+1)/2} g(x, q^{1/2}z)$$

and therefore

$$Yf_m = q^{(k+m)/2} f_m, \quad Yg_m = q^{-(k+m+1)/2} g_m.$$

for all integers $m \geq 0$.

From (6.2.2) it follows that f_m (resp. g_m) is a scalar multiple of E_{-m} (resp. E_{m+1}). Since the coefficients of x^{-m} in f_m and of x^{m+1} in g_m are each equal to $\begin{bmatrix} k+m \\ m \end{bmatrix}$, it follows from (6.2.5) and (6.2.6) that

(6.2.7) $$E_{-m} = \begin{bmatrix} k+m \\ m \end{bmatrix}^{-1} \sum_{i+j=m} \begin{bmatrix} k+i-1 \\ i \end{bmatrix} \begin{bmatrix} k+j \\ j \end{bmatrix} x^{i-j},$$

(6.2.8) $$E_{m+1} = \begin{bmatrix} k+m \\ m \end{bmatrix}^{-1} \sum_{i+j=m} \begin{bmatrix} k+i-1 \\ i \end{bmatrix} \begin{bmatrix} k+j \\ j \end{bmatrix} q^i x^{-i+j+1}$$

for all $m \geq 0$.

We shall now calculate $E_m(q^{-k/2})$, $m \in \mathbb{Z}$. We have

$$\begin{aligned} f(q^{-k/2}, z) &= 1/(q^{-k/2}z;q)_k \, (q^{k/2}z;q)_{k+1} \\ &= 1/(q^{-k/2}z;q)_{2k+1} \\ &= \sum_{m \geq 0} q^{-mk/2} \begin{bmatrix} 2k+m \\ m \end{bmatrix} z^m, \end{aligned}$$

and likewise

$$\begin{aligned} g(q^{-k/2}, z) &= q^{-k/2}/(q^{-k/2}z;q)_{k+1} (q^{(k+1)/2}z;q)_k \\ &= q^{-k/2}/(q^{-k/2}z;q)_{2k+1} \\ &= q^{-k/2} \sum_{m \geq 0} q^{-mk/2} \begin{bmatrix} 2k+m \\ m \end{bmatrix} z^m, \end{aligned}$$

from which it follows that

(6.2.9) $$\begin{aligned} E_{-m}(q^{-k/2}) &= q^{-mk/2} \begin{bmatrix} 2k+m \\ m \end{bmatrix} \Big/ \begin{bmatrix} k+m \\ m \end{bmatrix}, \\ E_{m+1}(q^{-k/2}) &= q^{-(m+1)k/2} \begin{bmatrix} 2k+m \\ m \end{bmatrix} \Big/ \begin{bmatrix} k+m \\ m \end{bmatrix}. \end{aligned}$$

As in §5.2, we can express xE_m and $x^{-1}E_m$ as linear combinations of the E_r. The formulas are

(6.2.10) $$xE_m = E_{m+1} - \frac{q^m(1-q^k)}{1-q^{m+k}} E_{1-m} \qquad (m \geq 1),$$

(6.2.11) $$xE_{-m} = \frac{(1-q^m)(1-q^{2k+m})}{(1-q^{k+m})^2} E_{1-m} + \frac{1-q^k}{1-q^{k+m}} E_{m+1} \qquad (m \geq 0),$$

(6.2.12) $$x^{-1}E_{1-m} = E_{-m} - \frac{1-q^k}{1-q^{m+k}} E_m \qquad (m \geq 1),$$

(6.2.13)
$$x^{-1}E_{m+1} = \frac{(1-q^m)(1-q^{2k+m})}{(1-q^{k+m})^2}E_m + \frac{q^m(1-q^k)}{1-q^{k+m}}E_{-m} \qquad (m \geq 0).$$

Proof These may be derived as in §5.2, or proved directly. To prove (6.2.13), for example, we observe that $(x^{-1}E_{m+1}, x^{m-2i}) = (E_{m+1}, x^{m+1-2i}) = 0$ for $1 \leq i \leq m$, so that $x^{-1}E_{m+1}$ is orthogonal to E_{m-2i} for $1 \leq i \leq m-1$. But $x^{-1}E_{m+1}$ is a linear combination of E_{m-2i} for $0 \leq i \leq m$, and since they are pairwise orthogonal it follows that

(1)
$$x^{-1}E_{m+1} = \lambda E_m + \mu E_{-m}$$

for scalars λ, μ to be determined. Here μ is the coefficient of x^{1-m} in E_{m+1}, which by (6.2.8) is $q^m(1-q^k)/(1-q^{k+m})$; and λ is determined by considering the coefficient of x^m on either side of (1), which gives

$$1 = \lambda + \mu(1-q^k)/(1-q^{k+m})$$

and hence the stated value for λ. □

From (6.2.13) we obtain

$$(E_{m+1}, E_{m+1}) = (E_{m+1}, x^{m+1}) = (x^{-1}E_{m+1}, x^m) = \frac{(1-q^m)(1-q^{2k+m})}{(1-q^{k+m})^2}(E_m, E_m)$$

for $m \geq 1$, since $(E_{-m}, E_m) = 0$. Hence

$$\frac{(E_m, E_m)}{(1,1)} = \prod_{i=1}^{m-1} \frac{(1-q^i)(1-q^{2k+i})}{(1-q^{k+i})^2}$$

and hence by (6.2.1)

(6.2.14)
$$(E_m, E_m) = \begin{bmatrix} 2k+m-1 \\ k \end{bmatrix} \Big/ \begin{bmatrix} k+m-1 \\ k \end{bmatrix}$$

for $m \geq 1$.

Now E_{m+1} and E_{-m} are related by

(6.2.15)
$$E_{m+1} = xE^*_{-m}$$

for $m \geq 0$; this follows from comparison of (6.2.7) and (6.2.8), or by simple considerations of orthogonality. Hence from (6.2.14)

(6.2.16)
$$(E_{-m}, E_{-m}) = \begin{bmatrix} 2k+m \\ k \end{bmatrix} \Big/ \begin{bmatrix} k+m \\ k \end{bmatrix}.$$

6.3 The symmetric polynomials P_m

Let

$$A_0 = \{f \in A : f(x) = f(x^{-1})\} = K[x + x^{-1}].$$

As in §5.1, the symmetric scalar product on A_0 is

$$<f, g> = <f, g>_k = \frac{1}{2}\mathrm{ct}(fg\nabla_k) \tag{6.3.1}$$

where $f, g \in A_0$ and

$$\nabla_k = (x^2; q)_k (x^{-2}; q)_k.$$

We shall assume as before that k is a non-negative integer when convenient. From (5.1.40) we have

$$<1, 1>_k = (1 + q^k)^{-1}(1, 1)_k$$

so that by (6.2.1)

$$<1, 1>_k = \begin{bmatrix} 2k-1 \\ k-1 \end{bmatrix}. \tag{6.3.2}$$

For each integer $m \geq 0$, let

$$P_m = P_{m,k} = P_{m\alpha/2,k}$$

in the notation of §5.3. By (5.3.3) the P_m are elements of A_0, pairwise orthogonal for the scalar product (6.3.1), and characterised by the facts that the coefficient of x^m in P_m is equal to 1, and that

$$(Y + Y^{-1})P_m = \left(q^{(m+k)/2} + q^{-(m+k)/2}\right) P_m. \tag{6.3.3}$$

Let $Z = (Y + Y^{-1}) \mid A_0$. Since $T_1 f = \tau f$ for $f \in A_0$, we have

$$\begin{aligned} Z &= \left(\tau + T_1^{-1}\right)U = (T_1 + \tau^{-1})U \\ &= (\boldsymbol{b}(X) + \tau^{-1})u + \boldsymbol{c}(X)s_1 u \\ &= \boldsymbol{c}(X^{-1})u + \boldsymbol{c}(X)s_1 u. \end{aligned} \tag{6.3.4}$$

To calculate the P_m, let z be an indeterminate and let

$$F(x, z) = F_k(x, z) = 1/(xz; q)_k(x^{-1}z; q)_k.$$

By the q-binomial theorem, the coefficient of z^m in $F(x, z)$ is

$$F_m = F_{m,k} = \sum_{i+j=m} \begin{bmatrix} k+i-1 \\ i \end{bmatrix} \begin{bmatrix} k+j-1 \\ j \end{bmatrix} x^{i-j}. \tag{6.3.5}$$

We now have

(6.3.6) $$ZF(x,z) = \tau F(x, q^{1/2}z) + \tau^{-1}F(x, q^{-1/2}z).$$

Proof From (6.3.4),

$$(x - x^{-1})ZF(x,z) = (\tau x - \tau^{-1}x^{-1})F(q^{1/2}x, z) + (\tau^{-1}x - \tau x^{-1})F(q^{-1/2}x, z)$$

On multiplying both sides by $q^{(k-1)/2}z(q^{-\frac{1}{2}}xz;q)_{k+1}(q^{-\frac{1}{2}}x^{-1}z;q)_{k+1}$, (6.3.6) is equivalent to

(1) $$(\delta - \gamma)\alpha\beta + (\beta - \alpha)\gamma\delta = (\beta - \gamma)\alpha\delta + (\delta - \alpha)\beta\gamma,$$

where $\alpha = (1 - q^{-1/2}xz)$, $\beta = (1 - q^{-1/2}x^{-1}z)$, $\gamma = (1 - q^{k-1/2}xz)$, $\delta = (1 - q^{k-1/2}x^{-1}z)$. Both sides of (1) are manifestly equal. □

From (6.3.6) it follows that

$$ZF_m = \left(q^{(k+m)/2} + q^{-(k+m)/2}\right)F_m$$

for each $m \geq 0$, and hence that F_m is a scalar multiple of P_m. Hence, from (6.3.5), we have

(6.3.7) $$P_m = \begin{bmatrix} k+m-1 \\ m \end{bmatrix}^{-1} \sum_{i+j=m} \begin{bmatrix} k+i-1 \\ i \end{bmatrix}\begin{bmatrix} k+j-1 \\ j \end{bmatrix} x^{i-j}.$$

Next, we have

$$F(q^{k/2}, z) = 1/(q^{-k/2}z;q)_{2k}$$

so that

$$F_m(q^{k/2}) = q^{-mk/2}\begin{bmatrix} 2k+m-1 \\ m \end{bmatrix}$$

and therefore

(6.3.8) $$P_m(q^{k/2}) = q^{-mk/2}\begin{bmatrix} 2k+m-1 \\ m \end{bmatrix} \Big/ \begin{bmatrix} k+m-1 \\ m \end{bmatrix} = q^{-mk/2}\prod_{i=0}^{m-1}\frac{1-q^{2k+i}}{1-q^{k+i}}.$$

The polynomials F_m are the continuous q-ultraspherical polynomials, in the terminology of [G1]. They were first introduced by L. J. Rogers in the 1890's [R1,2,3]. Precisely,

$$C_m(\cos\theta; q^k \mid q) = F_m(e^{i\theta})$$

in the notation of [G1], p. 169.

The norm formula (5.8.17) in the present case takes the form

$$\text{(6.3.9)} \qquad <P_m, P_m> = \prod_{i=1}^{k-1} \frac{1-q^{m+k+i}}{1-q^{m+k-i}} = \begin{bmatrix} 2k+m-1 \\ k \end{bmatrix} \Big/ \begin{bmatrix} m+k \\ k \end{bmatrix}.$$

In terms of the E's, we have

$$\text{(6.3.10)} \qquad P_m = E_{-m} + \frac{q^k(1-q^m)}{1-q^{k+m}} E_m.$$

Finally, we shall consider the shift operator (§5.9) in the present situation. Let

$$\delta_k(x) = \tau x - \tau^{-1} x^{-1}.$$

Then $\delta_k(Y^{-1}) = \tau T_1^{-1} U - \tau^{-1} U T_1$, so that for $f \in A_0$

$$\begin{aligned} \delta_k(Y^{-1}) f &= \left(\tau T_1^{-1} - 1\right) U f \\ &= \tau c(X)(s_1 - 1) u f. \end{aligned}$$

The shift operator is

$$G = \delta_k(X)^{-1} \delta_k(Y^{-1})$$

so that

$$Gf = \tau(x - x^{-1})^{-1}(s_1 u - u s_1) f$$

for $f \in A_0$. Apart from the factor $\tau = q^{k/2}$, this is independent of k, so we define

$$\text{(6.3.11)} \qquad G' = \tau^{-1} G = (x - x^{-1})^{-1}(s_1 u - u s_1)$$

as an operator on A_0. Explicitly,

$$(G'f)(x) = (x - x^{-1})^{-1}\left(f\left(q^{-1/2}x\right) - f\left(q^{-1/2}x\right)\right)$$

for $f \in A_0$. We calculate that

$$G' F_k(x, z) = q^{-1/2}(1 - q^k) z F_{k+1}\left(x, q^{-1/2} z\right)$$

so that

$$G'F_{m,k} = q^{-m/2}(1-q^k)F_{m-1,k+1}$$

and therefore

$$G'P_{m,k} = \left(q^{-m/2} - q^{m/2}\right)P_{m-1,k+1}. \tag{6.3.12}$$

It follows from (6.3.12) that

$$<G'P_{m,k}, P_{n-1,k+1}>_{k+1} = 0 \tag{6.3.13}$$

for $m \neq n$.

Let

$$\theta = s_1 u - u s_1, \quad \Phi_{k+1} = (x - x^{-1})^{-1}\nabla_{k+1}.$$

Then (6.3.13) takes the form

$$\mathrm{ct}(\theta(P_{m,k})\Phi_{k+1}P_{n-1,k+1}) = 0$$

or equivalently

$$\mathrm{ct}(P_{m,k}\theta(\Phi_{k+1}P_{n-1,k+1})) = 0$$

for $n \neq m$. This shows that $P_{m,k}$ is a scalar multiple of

$$\nabla_k^{-1}\theta(\Phi_{k+1}P_{m-1,k+1}) = \Phi_k^{-1}G'(\Phi_{k+1}P_{m-1,k+1}).$$

Consideration of the coefficient of x^{m+2k} in $\nabla_k P_{m,k}$ and in $\theta(\Phi_{k+1}P_{m-1,k+1})$ now shows that

$$P_{m,k} = \frac{q^{m/2}}{1-q^{2k+m}}\Phi_k^{-1}G'(\Phi_{k+1}P_{m-1,k+1}) \tag{6.3.14}$$

(and hence that $\Phi_k(X)^{-1} \circ G' \circ \Phi_{k+1}(X)$ is a scalar multiple of the shift operator $\hat{G} = \delta_k(Y)\delta_k(X)$).

Iterating (6.3.14) m times, we obtain

$$P_{m,k} = c_{m,k}\Phi_k^{-1}G'^m(\Phi_{k+m}), \tag{6.3.15}$$

where

$$c_{m,k} = q^{m(m+1)/4}/(q^{2k+m};q)_m \tag{6.3.16}$$

("Rodrigues formula").

6.4 Braid group and Hecke algebra (type $(C_1^\vee, C_1)$)

Suppose now that $S = S'$ is of type $(C_1^\vee, C_1)$, so that $R = R' = \{\pm\alpha\}$, where $|\alpha|^2 = 2$, and $L = L' = \mathbb{Z}\alpha$. The simple affine roots are $a_0 = \frac{1}{2} - \alpha$ and $a_1 = \alpha$, acting on $\mathbb{R}$: $a_0(\xi) = \frac{1}{2} - \xi$ and $a_1(\xi) = \xi$ for $\xi \in \mathbb{R}$. The simplex C is now the interval $(0, \frac{1}{2})$, and $W = W_S$ is the infinite dihedral group, generated by s_0 and s_1 where $s_0(\xi) = 1 - \xi$ and $s_1(\xi) = -\xi$.

The *braid group* $\mathfrak{B}$ is now the free group on two generators T_0, T_1. Let $Y = Y^\alpha$, so that

$$Y = T_0 T_1,$$

and $\mathfrak{B}$ is freely generated by T_1 and Y.

The *double braid group* $\tilde{\mathfrak{B}}$ is generated by T_1, X, Y, and a central element $q^{1/2}$, where $X = X^\alpha$. Since $<L, \alpha> = 2\mathbb{Z}$ the relations (3.4.1)–(3.4.5) are absent, i.e., there are no relations between T_1, X, Y, and $\tilde{\mathfrak{B}} \cong \mathbb{Z} \times F_3$ where F_3 is a free group on three generators. The antiautomorphism $\omega : \tilde{\mathfrak{B}} \to \tilde{\mathfrak{B}}$ is defined by

$$\omega\big(q^{1/2}, T_1, X, Y\big) = \big(q^{1/2}, T_1, Y^{-1}, X^{-1}\big).$$

Let

(6.4.1) $$T_0' = q^{-1/2} X T_0^{-1} = q^{-1/2} X T_1 Y^{-1}, \quad T_1' = X^{-1} T_1^{-1}.$$

Then we have

(6.4.2) $$T_0' T_0 T_1' T_1 = q^{-1/2},$$

and

(6.4.3) $$\omega(T_0, T_0', T_1, T_1') = (X T_1' X^{-1}, T_0', T_1, Y^{-1} T_0 Y).$$

Let $k = (k_1, k_2, k_3, k_4)$ be a labelling of S, as in §1.5, and let $k' = (k_1', k_2', k_3', k_4')$ be the dual labelling. Let

(6.4.4) $$\begin{aligned} \kappa_1 &= k_1 + k_2 = k_1' + k_2', & \kappa_1' &= k_1 - k_2 = k_3' + k_4', \\ \kappa_0 &= k_3 + k_4 = k_1' - k_2', & \kappa_0' &= k_3 - k_4 = k_3' - k_4', \end{aligned}$$

and let

(6.4.5) $$\tau_i = q^{\kappa_i/2}, \quad \tau_i' = q^{\kappa_i'/2} \qquad (i = 0, 1).$$

Thus replacing k by k' amounts to interchanging τ_0 and τ_1'.

Let $K = \mathbb{Q}(q^{1/2}, \tau_0, \tau_0', \tau_1, \tau_1')$ and let $A = K[x, x^{-1}]$, where $x = e^\alpha$. The *affine Hecke algebra* $\mathfrak{H}$ is the K-algebra generated by T_0 and T_1

subject to the relations

$$(T_i - \tau_i)(T_i + \tau_i^{-1}) = 0 \qquad (i = 0, 1). \tag{6.4.6}$$

As in §4.2, let

$$\boldsymbol{b}_i(x) = \boldsymbol{b}(\tau_i, \tau_i'; x) = \frac{\tau_i - \tau_i^{-1} + (\tau_i' - \tau_i'^{-1})x}{1 - x^2},$$

$$\boldsymbol{c}_i(x) = \boldsymbol{c}(\tau_i, \tau_i'; x) = \frac{\tau_i x - \tau_i^{-1}x^{-1} + \tau_i' - \tau_i'^{-1}}{x - x^{-1}}$$

for $i = 0, 1$. Then $\mathfrak{H}$ acts on A as follows:

$$T_i f = (\boldsymbol{b}_i(X_i) + \boldsymbol{c}_i(X_i)s_i)f \tag{6.4.7}$$

for $f \in A$, where $X_1 = X$ and $X_0 = q^{1/2}X^{-1}$, and

$$Xf = xf, \quad (s_0 f)(x) = f(qx^{-1}), \quad (s_1 f)(x) = f(x^{-1}).$$

The *double affine Hecke algebra* $\tilde{\mathfrak{H}}$ is generated over K by T_1, X, Y subject to the relations (6.4.6) and

$$(T_i' - \tau_i')(T_i' + \tau_i'^{-1}) = 0 \qquad (i = 0, 1) \tag{6.4.8}$$

where T_0', T_1' are given by (6.4.1). More symmetrically, $\tilde{\mathfrak{H}}$ is generated over K by T_0, T_0', T_1, T_1' subject to the relations (6.4.2), (6.4.6) and (6.4.8).

Dually, $\tilde{\mathfrak{H}}'$ has generators T_1, X, Y subject to the relations derived from (6.4.6) and (6.4.8) by interchanging τ_0 and τ_1'. Since by (6.4.3) $\omega(T_0)$ (resp. $\omega(T_1')$) is conjugate in $\tilde{\mathfrak{B}}$ to T_1' (resp. T_0), it follows that ω extends to an anti-isomorphism of $\tilde{\mathfrak{H}}'$ onto $\tilde{\mathfrak{H}}$.

6.5 The symmetric polynomials P_m

In the present case it is more convenient to consider the symmetric polynomials P_λ before the non-symmetric E_λ. As in §6.3, let $A_0 = K[x + x^{-1}]$. The symmetric scalar product on A_0 is now

$$<f, g> = <f, g>_k = \frac{1}{2}\mathrm{ct}(fg\nabla_k) \tag{6.5.1}$$

where now, as in (5.1.25),

$$\nabla_k = \frac{(x^2; q)_\infty (x^{-2}; q)_\infty}{\prod_{i=1}^{4} (u_i x; q)_\infty (u_i x^{-1}; q)_\infty}$$

and

(6.5.2) $$(u_1, u_2, u_3, u_4) = \left(q^{k_1}, -q^{k_2}, q^{1/2+k_3}, -q^{1/2+k_4}\right).$$

For each integer $m \geq 0$, let

$$P_m = P_{m,k} = P_{m\alpha,k}$$

in the notation of §5.3. The polynomials P_m are elements of A_0, pairwise orthogonal for the scalar product (6.5.1), and are characterized by the facts that P_m is a linear combination of $x^r + x^{-r}$ for $0 \leq r \leq m$, and that the coefficient of $x^m + x^{-m}$ is equal to 1. We have

(6.5.3) $$(Y + Y^{-1})P_m = (q^{m+k'_1} + q^{-m-k'_1})P_m.$$

Let $Z = (Y + Y^{-1}) \mid A_0$. Since $T_1 f = \tau_1 f$ for $f \in A_0$, we have

$$\begin{aligned} Z &= \tau_1 T_0 + T_1^{-1} T_0^{-1} \\ &= \tau_1 T_0 + \left(T_1 - \tau_1 + \tau_1^{-1}\right)\left(T_0 - \tau_0 + \tau_0^{-1}\right) \\ &= \left(T_1 + \tau_1^{-1}\right)(T_0 - \tau_0) + \tau_0\tau_1 + \tau_0^{-1}\tau_1^{-1} \\ &= (s_1 + 1)\boldsymbol{c}_1\left(X_1^{-1}\right)\boldsymbol{c}_0(X_0)(s_0 - 1) + q^{k'_1} + q^{-k'_1}. \end{aligned}$$

Now

$$\tau_1 \boldsymbol{c}_1\left(X_1^{-1}\right) = (1 - q^{k_1}X^{-1})(1 + q^{k_2}X^{-1})/(1 - X^{-2})$$

and

$$\tau_0 \boldsymbol{c}_0(X_0) = \left(1 - q^{k_3+1/2}X^{-1}\right)\left(1 + q^{k_4+1/2}X^{-1}\right)/(1 - qX^{-2}),$$

so that

(6.5.4) $$Z' = q^{k'_1} Z = (s_1 + 1) f(X^{-1})(s_0 - 1) + 1 + q^{2k'_1}$$

where

(6.5.5) $$f(x) = \left(\prod_{i=1}^{4}(1 - u_i x)\right) \Big/ (1 - x^2)(1 - qx^2).$$

To calculate the polynomials P_m explicitly, we shall use the symmetric polynomials

$$g_m(x) = (u_1 x; q)_m (u_1 x^{-1}; q)_m \qquad (m \geq 0)$$

as building blocks. They form a K-basis of A_0.

We have

(6.5.6) $$Z' g_m = \lambda_m g_m + \mu_m g_{m-1},$$

where

$$\lambda_m = q^{-m} + q^{m+2k'_1},$$
$$\mu_m = (1 - q^{-m})(1 - q^{m-1}u_1u_2)(1 - q^{m-1}u_1u_3)(1 - q^{m-1}u_1u_4).$$

Proof Since $s_0 x = qx^{-1}$ we calculate that

$$\frac{(s_0 - 1)g_m(x)}{1 - qx^{-2}} = q^{-1}u_1 x(q^m - 1)(u_1 x; q)_{m-1}(qu_1 x^{-1}; q)_{m-1}$$

and hence that

$$f(X^{-1})(s_0 - 1)g_m(x) = q^{-1}(q^m - 1)u_1 h(x) g_{m-1}(x),$$

where

$$h(x) = x^2(1 - q^{m-1}u_1 x^{-1})(1 - u_2 x^{-1})(1 - u_3 x^{-1})(1 - u_4 x^{-1})/(x - x^{-1}).$$

Now we have

$$\begin{aligned} h(x) + h(x^{-1}) = q^{1-m}u_1^{-1}&(1 - q^{m-1}u_1u_2)(1 - q^{m-1}u_1u_3)(1 - q^{m-1}u_1u_4) \\ &- q^{1-m}u_1^{-1}(1 - q^{2k'_1+m})(1 - q^{m-1}u_1 x)(1 - q^{m-1}u_1 x^{-1}). \end{aligned}$$

For both sides are linear in $x + x^{-1}$ and agree when $x = q^{m-1}u_1$, hence are proportional; moreover the coefficients of $x + x^{-1}$ on either side are equal to $1 - q^{2k'_1+m}$, since $u_1u_2u_3u_4 = q^{2k'_1+1}$.

Hence

$$\begin{aligned} Z'g_m &= (1 - q^{-m})(1 - q^{m-1}u_1u_2)(1 - q^{m-1}u_1u_3)(1 - q^{m-1}u_1u_4)g_{m-1} \\ &\quad + (1 + q^{2k'_1} - (1 - q^{-m})(1 - q^{2k'_1+m}))g_m \\ &= \lambda_m g_m + \mu_m g_{m-1}. \end{aligned}$$ □

Since the g_m form a K-basis of A_0, P_m is of the form

$$P_m = \sum_{r=0}^{m} \alpha_r g_r.$$

Hence by (6.5.3)

(1) $$Z'P_m = \lambda_m P_m = \sum_{r=0}^{m} \lambda_m \alpha_r g_r,$$

and by (6.5.6)

(2) $$Z'P_m = \sum_{r=0}^{m} \alpha_r(\lambda_r g_r + \mu_r g_{r-1})$$

(where $g_{-1} = 0$). From (1) and (2) we obtain

$$\lambda_m \alpha_r = \lambda_r \alpha_r + \mu_{r+1}\alpha_{r+1},$$

so that

$$\begin{aligned}\alpha_{r+1}/\alpha_r &= (\lambda_m - \lambda_r)/\mu_{r+1}\\ &= \frac{q(1-q^{-m+r})(1-q^{2k_1'+m+r})}{(1-q^{r+1})(1-q^r u_1u_2)(1-q^r u_1u_3)(1-q^r u_1u_4)},\end{aligned}$$

from which it follows that $P_{m,k}$ is a scalar multiple of the q-hypergeometric series

$$(6.5.7) \qquad \varphi_{m,k} = \sum_{r=0}^{m} \frac{q^r(q^{-m};q)_r(q^{2k_1'+m};q)_r(u_1x;q)_r(u_1x^{-1};q)_r}{(q;q)_r(u_1u_2;q)_r(u_1u_3;q)_r(u_1u_4;q)_r}$$

and more precisely that

$$(6.5.8) \qquad P_{m,k} = c_{m,k}\,\varphi_{m,k}$$

where

$$c_{m,k} = u_1^{-m}(u_1u_2;q)_m(u_1u_3;q)_m(u_1u_4;q)_m/(q^{2k_1'+m};q)_m.$$

The polynomials $P_{m,k}$ are (up to a scalar factor) the Askey-Wilson polynomials [A2]. Since ∇_k is symmetric in u_1, u_2, u_3, u_4, so are the $P_{m,k}$.

From (6.5.7) we have

$$\varphi_{m,k}(q^{k_1}) = \varphi_{m,k}(u_1) = 1$$

so that

$$(6.5.9) \qquad \varphi_{m,k} = \tilde{P}_{m,k}$$

in the notation of §5.3. Also from (6.5.7) we have

$$\begin{aligned}\varphi_{m,k}(q^{n+k_1}) &= \sum_{r\geq 0} \frac{q^r(q^{-m};q)_r(q^{-n};q)_r(q^{2k_1+n};q)_r(q^{2k_1'+m};q)_r}{(q;q)_r(u_1u_2;q)_r(u_1u_3;q)_r(u_1u_4;q)_r}\\ &= \varphi_{n,k'}(q^{m+k_1'})\end{aligned}$$

since $k_1 + k_i = k_1' + k_i'$ for $i = 2, 3, 4$. Hence

$$(6.5.10) \qquad \tilde{P}_{m,k}(q^{n+k_1}) = \tilde{P}_{n,k'}(q^{m+k_1'})$$

which is the symmetry law (5.3.5) in the present context.

The norm formula (5.8.17) in the present case, when expressed in terms of $u_1, \dots, u_4$ and $q^{2k_1'} = q^{-1}u_1u_2u_3u_4$, takes the form

$$(6.5.11)\qquad <P_m, P_m> = \frac{(q^{2m+2k_1'};q)_\infty (q^{2m+2k_1'+1};q)_\infty}{(q^{m+1};q)_\infty (q^{m+2k_1'};q)_\infty \prod_{i<j}(q^m u_i u_j;q)_\infty}.$$

Let G' be the operator on A_0 defined by (6.3.11). Then we have

$$(6.5.12)\qquad G' P_{m,k} = (q^{-m/2} - q^{m/2}) P_{m-1,k+1/2}.$$

Proof We calculate that

$$G' g_r = u_1 q^{-1/2}(q^r - 1)\left(u_1 q^{1/2} x;q\right)_{r-1}\left(u_1 q^{1/2} x^{-1};q\right)_{r-1}$$

from which and (6.5.7) it follows that $G'\varphi_{m,k}$ is a scalar multiple of $\varphi_{m-1,k+1/2}$, and hence that $G'P_{m,k}$ is a scalar multiple of $P_{m-1,k+1/2}$, where

$$k + \tfrac{1}{2} = (k_1 + \tfrac{1}{2}, k_2 + \tfrac{1}{2}, k_3 + \tfrac{1}{2}, k_4 + \tfrac{1}{2}).$$

Since

$$G'(x^m + x^{-m}) = \left(q^{-m/2} - q^{m/2}\right)\left(x^{m-1} + \cdots + x^{1-m}\right)$$

we have (6.5.12). □

Let

$$\Phi_k = \nabla_k/(x - x^{-1}).$$

Then

$$(6.5.13)\qquad P_{m,k} = c_{m,k}\Phi_k^{-1} G'\left(\Phi_{k+\frac{1}{2}} P_{m-1,k+\frac{1}{2}}\right)$$

where

$$c_{m,k} = -q^{m/2}/(1 - q^{2k_1'+m}).$$

Proof From (6.5.12) it follows that

$$<G' P_{m,k}, P_{n-1,k+\frac{1}{2}}>_{k+\frac{1}{2}} = 0$$

whenever $m \neq n$, or equivalently

$$(1)\qquad \mathrm{ct}\left(\theta(P_{m,k})\Phi_{k+\frac{1}{2}} P_{n-1,k+\frac{1}{2}}\right) = 0$$

where

$$(\theta f)(x) = f\left(q^{-1/2}x\right) - f\left(q^{1/2}x\right).$$

We may replace (1) by

$$\mathrm{ct}\big(P_{m,k}\theta\big(\Phi_{k+1/2}P_{n-1,k+1/2}\big)\big) = 0$$

i.e., by

$$<P_{m,k}, \Phi_k^{-1}G'\big(\Phi_{k+1/2}P_{n-1,k+1/2}\big)>_k = 0$$

whenever $m \neq n$. Hence $\Phi_k^{-1}G'(\Phi_{k+1/2}P_{m-1,k+1/2})$ is a scalar multiple of $P_{m,k}$.

Now

$$\Phi_{k+1/2}\big(q^{1/2}x\big)/\Phi_k(x) = -q^{-1/2}x^{-2}\prod_{i=1}^{4}(1-u_ix) = -q^{1/2}u(x),$$

say; and since $\Phi_k(x^{-1}) = -\Phi_k(x)$ we have

$$\Phi_{k+1/2}\big(q^{-1/2}x\big)/\Phi_k(x) = -q^{-1/2}u(x^{-1}),$$

so that

(2) $$\Phi_k^{-1}G'\big(\Phi_{k+1/2}P_{m-1,k+1/2}\big) = q^{1/2}(p(x)-p(x^{-1}))/(x-x^{-1}),$$

where

$$p(x) = u(x)P_{m-1,k+1/2}\big(q^{1/2}x\big).$$

Since

$$u(x) = q^{2k_1'+1}x^2 + \cdots + x^{-2}.$$

we have

$$p(x) = q^{2k_1'+(m+1)/2}x^{m+1} + \cdots + q^{(1-m)/2}x^{-m-1}$$

and therefore the coefficient of $x^m + x^{-m}$ in (2) is $q^{-m/2}(q^{2k_1'+m}-1)$. □

From (6.5.13) it follows that

(6.5.14) $$P_{m,k} = d_{m,k}\Phi_k^{-1}G'^m\big(\Phi_{k+m/2}\big),$$

where

$$d_{m,k} = (-1)^m q^{m(m+1)/4}/(q^{2k_1'+m};q)m,$$

(since when k is replaced by $k+\frac{1}{2}$, k_1' is replaced by $k_1'+1$).

6.6 The polynomials E_m

To calculate the polynomials $E_m = E_{m\alpha}$ ($m \in \mathbb{Z}$) we proceed as follows. The symmetric polynomials P_m were defined with reference to $x_0 = 0$ as origin; they are stable under s_1 (so that $T_1 P_m = \tau_1 P_m$) and are eigenfunctions of the operator $Z = T_0 T_1 + T_1^{-1} T_0^{-1}$, with eigenvalues $q^{m+k_1'} + q^{-m-k_1'}$. Equally, they are eigenfunctions of the operator

$$Z^\dagger = T_1 Z T_1^{-1} = T_1 T_0 + T_0^{-1} T_1^{-1}$$

on A with the same eigenvalues. If we now take $x_1 = \frac{1}{2}$ as origin, the effect is to interchange a_0 and a_1, T_0 and T_1, k_1 and k_3, and k_2 and k_4, so that the labelling k is replaced by

(6.6.1) $$k^\dagger = (k_3, k_4, k_1, k_2).$$

We obtain polynomials

(6.6.2) $$P_{m,k}^\dagger(x) = P_{m,k^\dagger}\left(q^{1/2} x^{-1}\right) = P_{m,k^\dagger}\left(q^{-1/2} x\right),$$

stable under s_0 (so that $T_0 P_m^\dagger = \tau_0 P_m^\dagger$) which are eigenfunctions of $Z^\dagger$ and of Z with eigenvalues $q^{m+k_1'} + q^{-m-k_1'}$.

In $P_m^\dagger$, $(u_1, \ldots, u_4)$ is replaced by

$$\left(q^{-1/2} u_3, q^{-1/2} u_4, q^{1/2} u_1, q^{1/2} u_2\right),$$

or equivalently, since $P_m^\dagger$ is symmetric in these four arguments, by

$$\left(q^{1/2} u_1, q^{1/2} u_2, q^{-1/2} u_3, q^{-1/2} u_4\right).$$

Hence by (6.5.7) and (6.5.8) we have

(6.6.3) $$P_{m,k}^\dagger = c_{m,k}^\dagger \varphi_{m,k}^\dagger$$

where

(6.6.4) $$\varphi_{m,k}^\dagger = \sum_{r=0}^{m} \frac{q^r (q^{-m}; q)_r (q^{2k_1'+m}; q)_r (u_1 x; q)_r (q u_1 x^{-1}; q)_r}{(q; q)_r (q u_1 u_2; q)_r (u_1 u_3; q)_r (u_1 u_4; q)_r}$$

and

(6.6.5) $$c_{m,k}^\dagger = \frac{q^{-m/2} u_1^{-m} (q u_1 u_2; q)_m (u_1 u_3; q)_m (u_1 u_4; q)_m}{(q^{2k_1'+m}; q)_m}$$
$$= q^{-m/2} \frac{1 - q^m u_1 u_2}{1 - u_1 u_2} c_{m,k}.$$

Now for each $m \geq 0$ the space

$$V_m = \left\{ f \in A : Zf = (q^{m+k_1'} + q^{-m-k_1'}) f \right\}$$

is two-dimensional, spanned by E_m and E_{-m}. From above, V_m is also spanned by P_m and $P_m^\dagger$. Hence each of E_m, E_{-m} is a linear combination of P_m and $P_m^\dagger$. Since

$$P_m = x^m + x^{-m} + \cdots, \quad P_m^\dagger = q^{-m/2} x^m + q^{m/2} x^{-m} + \cdots,$$

and since E_m does not contain x^{-m}, it follows that

(6.6.6) $$E_m = \frac{P_m - q^{-m/2} P_m^\dagger}{1 - q^{-m}}.$$

Next, we have by (5.7.8)

$$P_m = \lambda E_m + E_{-m}$$

where

$$\begin{aligned} \lambda &= \tau_1 \boldsymbol{c}\left(\tau_1, \tau_0; q^{-(m+k_1')}\right) \\ &= \frac{(1 - q^{-m})(1 + q^{-m-k_1'+k_2'})}{1 - q^{-2m-2k_1'}} \\ &= -\frac{(1 - q^m)(u_1 u_2 - q^{m+2k_1'})}{1 - q^{2m+2k_1'}}. \end{aligned}$$

From this and (6.6.6) we obtain

(6.6.7) $$E_{-m} = \frac{1 - q^m u_1 u_2}{1 - q^{2m+2k_1'}} P_m + \frac{q^{m/2}(u_1 u_2 - q^{m+2k_1'})}{1 - q^{2m+2k_1'}} P_m^\dagger$$

for $m \geq 0$. These two formulas (6.6.6) and (6.6.7) give E_m and E_{-m} explicitly as linear combinations of the two q-hypergeometric series $\varphi_{m,k}^\dagger$ (6.6.4) and $\varphi_{m,k}$ (6.5.7). Namely:

(6.6.8) *For all $m \in \mathbb{Z}$, E_m is a scalar multiple of*

$$(1 - u_1 u_2)\varphi_{|m|,k} + (u_1 u_2 - q^{k_1' - \bar{m}})\varphi_{|m|,k}^\dagger,$$

where

$$\bar{m} = \begin{cases} m + k_1' & \text{if } m > 0, \\ -m - k_1' & \text{if } m \leq 0. \end{cases}$$

This follows from (6.6.5), (6.6.6) and (6.6.7). □

Finally, we consider creation operators for the E_m. From §5.2, the E_m are the eigenfunctions of the operator Y on $A = K[x, x^{-1}]$:

(6.6.9)
$$YE_m = \begin{cases} q^{-m-k_1'}E_m & \text{if } m > 0, \\ q^{m+k_1'}E_m & \text{if } m \leq 0. \end{cases}$$

From (4.7.3) we have

$$(T_i - b_i'(X))X_i^{-1} = X_i(T_i - b_i'(X)) \qquad (i = 0, 1)$$

in $\tilde{\mathfrak{H}}'$, where $X_1 = X$, $X_0 = q^{1/2}X^{-1}$, and

$$b_0'(X) = b(\tau_1', \tau_0'; q^{1/2}X^{-1}),$$
$$b_1'(X) = b(\tau_1, \tau_0; X).$$

Applying $\omega\colon \tilde{\mathfrak{H}}' \to \tilde{\mathfrak{H}}$ gives

$$Y(T_1 - b(\tau_1, \tau_0; Y^{-1})) = (T_1 - b(\tau_1, \tau_0; Y^{-1}))Y^{-1}$$

and (since $\omega(T_0) = T_1^{-1}X^{-1}$)

$$q^{-1/2}Y^{-1}(T_1^{-1}X^{-1} - b(\tau_1', \tau_0'; q^{1/2}Y) = q^{1/2}(T_1^{-1}X^{-1} - b(\tau_1', \tau_0'; q^{1/2}Y).$$

So if we define

$$\alpha_0 = T_1^{-1}X^{-1} - b(\tau_1', \tau_0'; q^{1/2}Y),$$
$$\alpha_1 = T_1 - b(\tau_1, \tau_0; Y^{-1})$$

we have

(6.6.10)
$$Y\alpha_0 = q^{-1}\alpha_0 Y^{-1}, \quad Y\alpha_1 = \alpha_1 Y^{-1}.$$

The operators α_0, α_1 on A are 'creation operators': namely

(6.6.11) *We have*

$$\alpha_0 E_{-m} = \tau_1 E_{m+1} \qquad (m \geq 0),$$
$$\alpha_1 E_m = \tau_1^{-1} E_{-m} \qquad (m > 0).$$

Proof Consider $\alpha_1 E_m$. Since

$$Y\alpha_1 E_m = \alpha_1 Y^{-1} E_m = q^{m+k_1'}\alpha_1 E_m,$$

it follows from (6.6.9) that $\alpha_1 E_m$ is a scalar multiple of E_m. To find the scalar multiple, consider the coefficient of x^{-m} in $\alpha_1 E_m$. Now $b(\tau_1, \tau_0; Y^{-1})E_m$ is a scalar multiple of E_m, hence does not contain x^{-m}; also

$$T_1 x^m = \tau_1 x^m + \big(\tau_1 x - \tau_1^{-1} x^{-1} + \tau_0 - \tau_0^{-1}\big)(x^{-m} - x^m)/(x - x^{-1})$$

in which the coefficient of x^{-m} is τ_1^{-1}. Hence $\boldsymbol{\alpha}_1 E_m = \tau_1^{-1} E_{-m}$.

Next,

$$Y \boldsymbol{\alpha}_0 E_{-m} = q^{-1} \boldsymbol{\alpha}_0 Y^{-1} E_{-m} = q^{-(m+1+k_1')} \boldsymbol{\alpha}_0 E_{-m},$$

so that by (6.6.9) $\boldsymbol{\alpha}_0 E_{-m}$ is a scalar multiple of E_{m+1}. As before, $\boldsymbol{b}(\tau_1', \tau_0'; q^{1/2} Y)$ E_{-m} is a scalar multiple of E_{-m}, hence does not contain x^{m+1}; and

$$T_1^{-1} X^{-1} x^{-m} = \tau_1^{-1} x^{-m-1} + \big(\tau_1 x - \tau_1^{-1} x^{-1} + \tau_0 - \tau_0^{-1}\big)(x^{m+1} - x^{-m-1})/(x - x^{-1})$$

in which the coefficient of x^{m+1} is τ_1. Hence $\boldsymbol{\alpha}_0 E_{-m} = \tau_1 E_{m+1}$. □

From (6.6.11) it follows that

(6.6.12) $$E_{-m} = (\boldsymbol{\alpha}_1 \boldsymbol{\alpha}_0)^m (1)$$

for $m \geq 0$, and

(6.6.13) $$E_m = \tau_1^{-1} \boldsymbol{\alpha}_0 (\boldsymbol{\alpha}_1 \boldsymbol{\alpha}_0)^{m-1} (1)$$

for $m \geq 1$.

Bibliography

[A1] G. E. Andrews, Problems and prospects for basic hypergeometric functions. In *Theory and Applications of Special Functions*, ed. R. Askey, Academic Press, New York (1975).

[A2] R. Askey and J. Wilson, Some basic hypergeometric orthogonal polynomials that generalize Jacobi polynomials, *Memoirs of the American Mathematical Society* **319** (1985).

[B1] N. Bourbaki, *Groupes et algèbres de Lie*, Chapitres 4, 5 et 6, Hermann, Paris (1968).

[B2] E. Brieskorn and K. Saito, Artin-gruppen und Coxeter-gruppen, *Inv. Math.* **17** (1972) 245–271.

[B3] F. Bruhat and J. Tits, Groupes réductifs sur un corps local: I. Données radicielles valuées, *Publications Mathématiques de l'Institut des Hautes Études Scientifiques*, no. **41** (1972).

[C1] I. Cherednik, Double affine Hecke algebras, Knizhnik – Zamolodchikov equations, and Macdonald's operators, *International Mathematics Research Notices* **9** (1992). 171–179.

[C2] I. Cherednik, Double affine Hecke algebras and Macdonald's conjectures, *Ann. Math.* **141** (1995) 191–216.

[C3] I. Cherednik, Non-symmetric Macdonald polynomials, *International Mathematics Research Notices* **10** (1995) 483–515.

[C4] I. Cherednik, Macdonald's evaluation conjectures and difference Fourier transform, *Inv. Math.* **122** (1995) 119–145.

[C5] I. Cherednik, Intertwining operators of double affine Hecke algebras, *Sel. Math.* new series **3** (1997) 459–495.

[D1] F. J. Dyson, Statistical theory of the energy levels of complex systems I, *J. Math. Phys.* **3** (1962) 140–156.

[G1] G. Gasper and M. Rahman, *Basic Hypergeometric Series*, Cambridge University Press (1990).

[G2] R. A. Gustafson, A generalization of Selberg's beta integral, *Bulletin of the American Mathematical Society* **22** (1990) 97–105.

[H1] G. J. Heckman and E. M. Opdam, Root systems and hypergeometric functions I, *Comp. Math.* **64** (1987) 329–352.

[H2] G. J. Heckman, Root systems and hypergeometric functions II, *Comp. Math.* **64** (1987) 353–373.

[I1] B. Ion, Involutions of double affine Hecke algebras, preprint (2001).

[K1] V. G. Kac, *Infinite Dimensional Lie Algebras*, Birkhäuser, Boston (1983).

[K2] A. A. Kirillov, Lectures on affine Hecke algebras and Macdonald's conjectures, *Bulletin of the American Mathematical Society* **34** (1997) 251–292.

[K3] T. Koornwinder, Askey-Wilson polynomials for root systems of type BC, *Contemp. Math.* **138** (1992) 189–204.

[L1] G. Lusztig, Affine Hecke algebras and their graded version, *Journal of the American Mathematical Society* **2** (1989) 599–635.

[M1] I. G. Macdonald, Spherical functions on a group of p-adic type, *Publications of the Ramanujan Institute*, Madras (1971).

[M2] I. G. Macdonald, Affine root systems and Dedekind's η-function, *Inv. Math.* **15** (1972) 91–143.

[M3] I. G. Macdonald, The Poincaré series of a Coxeter group, *Math. Annalen* **199** (1972) 161–174.

[M4] I. G. Macdonald, Some conjectures for root systems, *SIAM Journal of Mathematical Analysis* **13** (1982) 988–1007.

[M5] I. G. Macdonald, Orthogonal polynomials associated with root systems, preprint (1987); *Séminaire Lotharingien de Combinatoire* **45** (2000) 1–40.

[M6] I. G. Macdonald, *Symmetric Functions and Hall Polynomials*, 2nd edition, Oxford University Press (1995).

[M7] I. G. Macdonald, Affine Hecke algebras and orthogonal polynomials, *Astérisque* **237** (1996) 189–207.

[M8] I. G. Macdonald, Symmetric functions and orthogonal polynomials, *University Lecture Series* vol. **12**, American Mathematical Society (1998).

[M9] R. V. Moody, A new class of Lie algebras, *J. Algebra* **10** (1968) 211–230.

[M10] R. V. Moody, Euclidean Lie algebras, *Can. J. Math.* **21** (1969) 1432–1454.

[M11] W. G. Morris, *Constant Term Identities for Finite and Affine Root Systems: Conjectures and Theorems*, Ph.D. thesis, Madison (1982).

[N1] M. Noumi, Macdonald – Koornwinder polynomials and affine Hecke rings, *Sūriseisekikenkyūsho Kōkyūroku* **919** (1995) 44–55 (in Japanese).

[O1] E. M. Opdam, Root systems and hypergeometric functions III, *Comp. Math.* **67** (1988) 21–49.

[O2] E. M. Opdam, Root systems and hypergeometric functions IV, *Comp. Math.* **67** (1988) 191–209.

[O3] E. M. Opdam, Some applications of hypergeometric shift operators, *Inv. Math.* **98** (1989) 1–18.

[O4] E. M. Opdam, Harmonic analysis for certain representations of graded Hecke algebras, *Acta Math.* **175** (1995) 75–121.

[R1] L. J. Rogers, On the expansion of some infinite products, *Proc. London Math. Soc.* **24** (1893) 337–352.

[R2] L. J. Rogers, Second memoir on the expansion of certain infinite products, *Proc. London Math. Soc.* **25** (1894) 318–343.

[R3] L. J. Rogers, Third memoir on the expansion of certain infinite products, *Proc. London Math. Soc.* **26** (1895) 15–32.

[S1] S. Sahi, Nonsymmetric Koornwinder polynomials and duality, *Arm. Math.* **150** (1999) 267–282.

[S2] S. Sahi, Some properties of Koornwinder polynomials, *Contemp. Math.* **254** (2000) 395–411.

[S3] R. Stanley, Some combinatorial properties of Jack symmetric functions, *Adv. in Math.* **77** (1989) 76–115.

[S4] J. V. Stokman, Koornwinder polynomials and affine Hecke algebras, preprint (2000).

[V1] H. van der Lek, *The Homotopy Type of Complex Hyperplane Arrangements*, Thesis, Nijmegen (1983).

[V2] J. van Diejen, Self-dual Koornwinder-Macdonald polynomials, *Inv. Math.* **126** (1996) 319–339.

Index of notation

Index